Lecture Notes in Computer Science 4936

Commenced Publication in 1973
Founding and Former Series Editors:
Gerhard Goos, Juris Hartmanis, and Jan van Leeuwen

T0223309

William Aiello   Andrei Broder
Jeannette Janssen   Evangelos Milios (Eds.)

Algorithms and Models for the Web-Graph

Fourth International Workshop, WAW 2006
Banff, Canada, November 30 - December 1, 2006
Revised Papers

 Springer

Volume Editors

William Aiello
University of British Columbia
Vancouver, Canada
E-mail: aiello@cs.ubc.ca

Andrei Broder
Yahoo! Research
Sunnyvale, CA 94089, USA
E-mail: broder@yahoo-inc.com

Jeannette Janssen
Dalhousie University
Halifax, B3H 3J5, Canada
E-mail: janssen@mathstat.dal.ca

Evangelos Milios
Dalhousie University
Halifax, B3H 1W5, Canada
E-mail: eem@cs.dal.ca

Library of Congress Control Number: 2008922735

CR Subject Classification (1998): H.5.1, H.5.3, H.5.4

LNCS Sublibrary: SL 3 – Information Systems and Application, incl. Internet/Web
and HCI

ISSN   0302-9743
ISBN   3-540-78807-7 Springer Berlin Heidelberg New York
ISBN   978-3-540-78807-2 Springer Berlin Heidelberg New York

Springer is a part of Springer Science+Business Media

springer.com

© Springer-Verlag Berlin Heidelberg 2008

Typesetting: Camera-ready by author, data conversion by Scientific Publishing Services, Chennai, India
Printed on acid-free paper   SPIN: 12246403   06/3180   5 4 3 2 1 0

Preface

The World Wide Web has become part of our everyday life and information retrieval and data mining on the Web are now of enormous practical interest. The algorithms supporting these activities combine the view of the Web as a text repository and as a graph, induced in various ways by links among pages, links among hosts, or other similar networks.

The aim of the 4th Workshop on Algorithms and Models for the Web-Graph (WAW 2006) was to further the understanding of these Web-induced graphs and stimulate the development of high-performance algorithms and applications that use the graph structure of the Web. The workshop was meant both to foster an exchange of ideas among the diverse set of researchers already involved in this topic and to act as an introduction for the larger community to the state of the art in this area. The workshop program included invited keynote talks by Fan Chung-Graham (UCSD), Soumen Chakrabarti (IITB), Walter Willinger (ATT Research) and Filippo Menczer (Indiana).

WAW 2006 took place on November 30 and December 1 at the Banff International Research Institute (BIRS), in Banff, Alberta (Canada). It was the fourth in a series of very successful workshops on the Web graph. WAW 2002 (Vancouver) and 2004 (Rome) were held in conjunction with the Annual IEEE Symposium on Foundations of Computer Science (FOCS). WAW 2003 (Budapest) was held in conjunction with the 12th International World Wide Web Conference.

In response to the call for abstracts we received 28 submissions. Almost all submissions were relevant to the topic of the workshop and contained interesting ideas. The Organizing Committee selected 12 contributors to present their work at the workshop, while 12 others were invited to prepare a poster about their work, which was exhibited at the workshop. Those authors selected for contributed talks were invited to prepare a paper for the proceedings. The poster presenters were given an opportunity to explain their work during a poster session. All participants were invited to judge the posters in a contest for best poster. The three posters that received the most votes were also invited to submit their work to the proceedings. The papers were then submitted to a refereeing process by the Program Committee, and the final versions are included in this volume. The volume also includes abstracts of the invited talks.

We would like to thank all those that helped to make this workshop a success, with special thanks to the staff at BIRS. We thank the MITACS (Mathematics of Information Technology and Complex Systems) Network of Centres of Excellence (NCE) for financial and organizational support. We also owe many thanks to Andrei Voronkov for providing the EasyChair conference system. With this system, managing the refereeing process and the collation of the proceedings became almost effortless. Finally, we would like to thank all participants in the workshop, all authors of the contributed papers, and especially the invited speakers for their contribution in making WAW 2006 an amicable and stimulating forum for the exchange of ideas about the Web graph.

November 2007

William Aiello
Andrei Broder
Jeannette Janssen
Evangelos Milios

Organization

Organizing Committee

William Aiello   University of British Columbia
Andrei Broder   Yahoo! Inc.
Jeannette Janssen   Dalhousie University
Evangelos Milios   Dalhousie University

Sponsoring Institutions

The Mathematics of Information Technology and Complex Systems (MITACS)
Yahoo! Inc.
GenieKnows.com

Table of Contents

Modelling and Mining of Networked Information Spaces

William Aiello[1], Andrei Broder[2], Jeannette Janssen[3], and Evangelos Milios[4]

[1] Department of Computer Science, University of British Columbia
[2] Yahoo! Research
[3] Department of Mathematics and Statistics, Dalhousie University
[4] Faculty of Computer Science, Dalhousie University
eem@cs.dal.ca

1 Overview

In recent years, the emergence of the Web and the dramatic increase in computing, storage and networking capacity has given rise to the concept of networked
information spaces. The prime example of a networked information space is
the World Wide Web itself. The Web, in its pure form, is a set of hypertext
documents, with links in one document pointing to another document. Other
networked information spaces enabled by and built on top of the Web include:

- *Online Case Law*, where judicial cases form precedents for similar cases which
  come after them. Judicial decisions are in effect laws that fill in gaps in
  statute law. As a result, a case makes reference to other cases or to statutes
  to support the judicial decision, leading the creation of a complex set of links
  present in cases.
- *The scientific literature*, where articles cite and are cited other articles, leading to the creation of a citation graph, where vertices are publications and
  directional edges are citations between articles. Co-authorship relations between authors give rise to a different graph, the co-authorship graph, where
  vertices are authors and edges are co-authorships of papers.
- The patent literature, which is interconnected in a manner similar to the
  scientific literature.
- *Blog space*, a recent extension of the World Wide Web, which significantly
  lowers the threshold of expertise required for an individual to post web content.
- *Social information spaces*, such as `del.icio.us`, where people share their
  bookmarks, which are tagged with keywords, and connect to each other via
  explicit friendship links.
- Collaboratively authored Web resources, such as the Wikipedia[1], YouTube[2],
  Facebook[3] and Flickr[4].

---

[1] http://en.wikipedia.org/wiki/Wikipedia
[2] http://www.youtube.com/
[3] http://www.facebook.com/
[4] http://www.flickr.com/

W. Aiello et al. (Eds.): WAW 2006, LNCS 4936, pp. 1–17, 2008.

A key characteristic of networked information spaces is their *social nature* and their *organic growth*. They are not the product of a single individual, but of an entire collaborating community, resulting in the creation of both comprehensive information resources and social networks. Networked information spaces represent both resources to be tapped into (through appropriate information search and retrieval tools), as well as entities that can be studied as physical phenomena (i.e. measured and modeled). Their massive scale and distributed implementation requires the development of both new technology and new scientific methods for their study. More specifically, open algorithmic problems are:

- Information Search and Retrieval. Open issues are (a) how to cope algorithmically with the massive scale of the information spaces, requiring ever increasing computational and storage resources, (b) how to rank search results and introduce personalization, where the personal context of the search is taken into account.
- Organization of the networked information space, which includes forming possibly hierarchical clusters of documents and authors, making browsing a complementary activity to search.

The science of networked information spaces requires the development of new mathematical models for describing their dynamic nature, in a manner that is consistent with the properties of real networks. Given the massive scale of the networks, appropriate abstraction mechanisms need to be developed forming the "lenses" through which to observe their dynamic behaviour, and thereby test the validity of the proposed mathematical models.

The MoMiNIS Winter School provided the opportunity to a group of selected graduate students to attend tutorials in the Modelling and Mining of Networked Information Spaces from recognized experts in the field, and present their own work and receive feedback. The unique setting of BIRS was highly conducive to personal interaction and generation of new ideas. The Winter School had the following objectives:

- To provide knowledge of the state of the art in the field via tutorials by experts
- To support young researchers with networking
- To provide young researchers with feedback on their research and the opportunity to discuss their research interests with established researchers in the field

2   Presentation Highlights

The Winter school consisted of four invited tutorials, a number of poster presentations, and a panel discussion based on a set of questions put to the panelists[5].

---

[5] In this section, we give a summary of the tutorials held at the workshop. Any perceived bias, error or oversight should be attributed not to the lecturers, but to the inevitable inaccuracies of the editors' observations.

2.1 Tutorials

On measuring, inferring, and modeling Internet connectivity: A guided tour across the TCP/IP protocol stack by Dr. Walter Willinger, *AT&T Labs-Research.*

Dr. Willinger's tutorial focused on the intricacies of measuring Internet topology, as seen from TCP/IP protocol stack. One of the most visible manifestations of the Internets vertical decomposition is the 5-layer TCP/IP protocol stack. This layered architecture gives rise to a number of different connectivity structures, with the lower layers (e.g., router-level) defining more physical and the higher layers (e.g., the Web) more virtual or logical types of topologies. The resulting graph structures have been designed with very different objectives in mind, have evolved according to different circumstances, and have been shaped by succinctly different forces. The main objective of this tutorial was to discuss the problems and challenges associated with measuring, inferring, and modeling these different connectivity structures. To this end, the tutorial was divided into the following four parts:

1. **Measurements:** Internet connectivity measurements are notorious for their ambiguities, inaccuracies, and incompleteness. As a general rule, they should never be taken at face value, but need to be scrutinized for consistency with the networking context from which they were obtained, and to do so, it is important to understand the process by which they were collected.

2. **Inference:** The challenge is to know whether or not the results we infer from our measurements are indeed well-justified claims, and at issue are the quality of the measurements themselves, the quality of their analysis, and the sensitivity of the inferred properties to known imperfections of the measurements.

3. **Modeling:** Developing appropriate models of Internet connectivity that elucidate observed structure or behavior is typically an under-constrained problem, meaning that there are in general many different explanations for one and the same phenomenon. To argue in favor of any particular explanation typically involves additional information, either in the form of domain knowledge or of new or complementary data. It is in the choice of this side information and how it is incorporated into the model building process, where considerable differences arise in the various approaches to Internet topology modeling that have been applied to date.

4. **Model validation:** There has been an increasing awareness of the fact that the ability to replicate some statistics of the original data or inferred quantities does not constitute validation for a particular model. While one can always use a model with enough parameters to fit a given data set, such models are merely descriptive and have in general no explanatory power. For the problems described here, appropriate validation typically means additional work (e.g., identifying and collecting complementary measurements that can be used to check a proposed explanation).

The take-home lesson from this tutorial was that measuring Internet connectivity is a non-trivial task, and the tools available cannot measure everything we are interested in measuring. As a result, we should be very careful about the claims we make about models based on incomplete measurements on the Internet.

Walter Willinger received the Diplom (Dipl. Math.) from the ETH Zurich, Switzerland, and the M.S. and Ph.D. degrees from the School of ORIE, Cornell University, Ithaca, NY. He is currently a member of the Information and Software Systems Research Center at AT&T Labs - Research, Florham Park, NJ, and before that, he was a Member of Technical Staff at Bellcore Applied Research (1986-1996). His research interests include studying the multiscale nature of Internet traffic and topology and developing a theoretical foundation for dealing with large-scale communication networks such as the Internet. He is a Fellow of ACM (2005) and a Fellow of IEEE (2005). For his work on the self-similar (fractal) nature of Internet traffic, he received the 1996 IEEE W.R.G. Baker Prize Award, the 1994 W.R. Bennett Prize Paper Award, and the 2005 ACM/SIGCOMM Test of Time Paper Award.

Web Mining, mapping, modeling and mingling by *Prof. Filippo Menczer*, Indiana State University

The Web is a complex self-organized system whose evolution and use is shaped by many concurrent social, cognitive, economic, and information phenomena. This tutorial described ongoing efforts to study the topological and dynamical properties of link, content, and semantic networks stemming from some of these forces. It was discussed what we think, what we know, what we can use regarding the structure, content, and use of the Web, and what the future of intelligent, cooperative Web search may bring.

Prof. Menczer reviewed the concept of semantic similarity based on link structure and his study using the Open Directory Project data. He stated that content alone is a poor predictor of semantic similarity. Making reference to the Heuristically Optimized Trade-offs (HOT) work to explain power laws [6], he suggested that it is useful to incorporate content into the growth models based on topology, pointing to his recent work on the Evolution of Document Networks, and how to extend growth models to account for lexical similarity [7].

He then went on to discuss social networking through bookmark sharing. Consumer-driven tagging represents a paradigm shift, where both text and links are author-driven. An open problem is the study of the properties of relatedness between pages bookmarked by the same person. Can this lead to a new notion of similarity between pages, if averaged over many persons?

Another issue of social networking is spamming. Is there a trust infrastructure that can be overlaid on social networks? A third issue is that for a social networking service to succeed, it needs a critical mass of users.

In peer-to-peer search, users collaborate to speed up and improve search results. The key question is how to route queries to the right peers, and how to combine the results. The opportunity exists to learn models of other users by observing the process of querying, leading to the representation of semantic communities.

Filippo Menczer is an associate professor of informatics and computer science, adjunct associate professor of physics, and a member of the cognitive science program at Indiana University, Bloomington. He holds a Laurea in Physics from the University of Rome and a Ph.D. in Computer Science and Cognitive Science from the University of California, San Diego. Dr. Menczer has been the recipient of Fulbright, Rotary Foundation, and NATO fellowships, and is a fellow-at-large of the Santa Fe Institute. His research is supported by a Career Award from the National Science Foundationon and focuses on Web, text, and data mining, Web intelligence, distributed information systems, social Web search, adaptive intelligent agents, complex systems and networks, and artificial life.

Ranking and labeling graphs: Analysis of links and node attributes by *Prof. Soumen Chakrabarti*, IITB, Mumbai

In this tutorial, mathematical techniques for ranking and labeling nodes in a graph were discussed, based on the link structure of the graph as well as attributes of the nodes. Ranking and labeling have obvious applications in Web search and page classification, but the range of applications is widening to finer-grained entity-relationship graphs where nodes represent entities like people, emails, papers, organizations and locations and edges represent relations like works-for, wrote, cited, is-located-in. Applications also include annotating unstructured and semistructured sources with type tags which can then be indexed for search. On the subject of ranking, the presentation started with a general discussion on learning to rank feature vectors, given training data, using a maximum margin formulation. Ranking nodes in graphs is based on the intuition that nodes should score highly if high-scoring instances link to it, leading to two approaches, Hyperlink induced topic search (HITS), and Pagerank. The connection between HITS and SVD/PCA was presented, followed by the random walk model on which Pagerank is based. The stability of ranking in HITS and Pagerank was introduced. The discussion of ranking concluded with the presentation of refinements of the basic models, such as Probabilistic HITS variants, and Personalized and Topic-sensitive Pagerank. On the subject of labelling, the problem of collective labelling of a large number of instances, whose labels are not independent, was presented. SVM-based training and a probabilistic view of Markov networks was discussed, leading to relaxation-based algorithms for inference.

Soumen Chakrabarti received his B.Tech in Computer Science from the Indian Institute of Technology, Kharagpur, in 1991 and his M.S. and Ph.D. in Computer Science from the University of California, Berkeley in 1992 and 1996. At Berkeley he worked on compilers and runtime systems for running scalable parallel scientific software on message passing multiprocessors.

He was a Research Staff Member at IBM Almaden Research Center from 1996 to 1999, where he worked on the Clever Web search project and led the Focused Crawling project.

In 1999 he joined the Department of Computer Science and Engineering at the Indian Institute of Technology, Bombay, where he has been an Associate professor since 2003.

In Spring 2004 he was Visiting Associate professor at Carnegie-Mellon University.

He has published in the WWW, SIGIR, SIGKDD, SIGMOD, VLDB, ICDE, SODA, STOC, SPAA and other conferences as well as Scientific American, IEEE Computer, VLDB and other journals. He holds eight US patents on Web-related inventions. He has served as technical advisor to search companies and vice-chair or program committee member for WWW, SIGIR, SIGKDD, VLDB, ICDE, SODA and other conferences, and guest editor or editorial board member for DMKD and TKDE journals. He is also author of a book on Web Mining.

His current research interests include integrating, searching, and mining text and graph data models, exploiting types and relations in search, and Web graph and popularity analysis.

Navigation and Evolution of Social Networks by *Prof. David Liben-Nowell*, Carleton College, Minnesota

In this tutorial, Prof. Liben-Nowell introduced some of the empirical observations of the structure of social networks, especially in comparison to the structure of the web. He then discussed a number of algorithmic topics arising in social networks, including the latent information contained in social networks (how much information about people is implicit in their connections?) and how to search social networks (can you find a short path to a target without global knowledge of the graph?).

In the first part of the tutorial, starting with Milgram's experiments (1967) of forming chains between Omahaians and a stockbroker in Sharon, MA, that led to the notion of six degrees of separation, he went on to discuss properties of a variety of social networks. The high-school friendships network [Moody 2001] displayed both a small diameter and a high clustering coefficient, with the Watts-Strogatz model (1998) of a rewired ring lattice trying to model this behaviour. He next discussed the notion of Greedy Routing in social networks search that goes beyond homophily (a person's friends tend to be similar to him/her), to capture the notion that the next step in seeking a route to the target is chosen to maximimize the similarity to the target. Required conditions for the greedy routing algorithm to work include having well-scattered friends (to reach faraway targets) as well as well-localized friends (to home in on nearby targets). Rank-based friendship aims to capture the notion of non-uniformly distributed populations. It is a model that assigns a probability that A is a friend of B inversely proportional to the the number of people closer to A than the distance from A to B. In real life, there are may ways to define distance. How are they to be combined? Relevant research was reviewed.

The second part of the tutorial dealt with the information content of social networks. If we focus on Milgram's experiments, we note that only 18 chains were completed out of 96. There is little data on why failed chains failed, but it implies that certain targets may be significantly harder to reach than others. More recent small-world experiments confirmed that most chains fail. The notion of friendship is messier than originally thought: there are systematic friendships (due to geographical or occupational proximity) and random friendships (due to a serendipitous encounter). Evolution through common friends (closing the triangle) is another possibility, that explains high clustering coefficients. The final topic of the second part is the introduction of the link prediction problem, as a means to evaluate the various models discussed earlier. The idea is to train the model using network data up to a certain time, and then try to predict what links will appear at subsequent times. Experimental results demonstrate that although performance of prediction is significantly better than random, a large fraction of predictions are still wrong. Furthermore, only 5-10% of the most similar people end up friends. Finally, not all social networks behave the same. There are unsystematic networks (where predictors perform the worst), simplistic networks (predictors are best at rank=1), popularity contest networks (preferential attachment is too good).

David Liben-Nowell is an assistant professor of computer science at Carleton College, Minnesota. He received his PhD in theoretical computer science from MIT's Computer Science and Artificial Intelligence Laboratory in 2005. His research interests include a variety of applications of the techniques of theoretical computer science to questions arising within and beyond computer science, with a focus on large-scale information networks and their evolution. David's research interests also include game theory, peer-to-peer computing, and computational biology. Prior to coming to MIT, David received a BA from Cornell and an MPhil from the University of Cambridge.

Graph Theory in the Information Age by *Prof. Fan Chung-Graham*, University of California, San Diego

Recently, graph theory has emerged as a primary tool for detecting numerous hidden structures in various information networks, including Web graphs, social networks, biological networks, or more generally, any graph representing relations in massive data sets. Thus, through examples of large sparse graphs in realistic networks, research in graph theory has been forging ahead into an exciting new direction. This tutorial gave an overview of the various graph theoretic techniques that can be used to attack problem related to real life networks, and gave an overview of challenging open problems in this emerging research area.

Fan Chung-Graham received a B.S. degree in mathematics from National Taiwan University in 1970 and a Ph.D. in mathematics from the University of Pennsylvania in 1974, after which she joined the technical staff of AT&T Bell Laboratories. From 1983 to 1991, she headed the Mathematics, Information Sciences and Operations Research Division at Bellcore. In 1991 she became a Bellcore Fellow. In 1993, she was the Class of 1965 Professor of Mathematics at the the University of Pennsylvania. Since 1998, she has been a Professor of Mathematics and Professor of Computer Science and Enginering at the University of California, San Diego. She is also the Akamai Professor in Internet Mathematics.

Her research interests are primarily in graph theory, combinatorics, and algorithmic design, in particular in spectral graph theory, extremal graphs, graph labeling, graph decompositions, random graphs, graph algorithms, parallel structures and various applications of graph theory in Internet computing, communication networks, software reliability, chemistry, engineering, and various areas of mathematics. She was awarded the Allendoerfer Award by Mathematical Association of America in 1990. Since 1998, she has been a member of the American Academy of Arts and Sciences.

2.2   Panel Discussion

The winter school concluded with a panel discussion, loosely based on a list of questions posed to the panelists by the workshop organizers. The panel was moderated by Bill Aiello and the panelists were Andrei Broder, David Liben-Nowell, Fan Chung, Filippo Menczer, Soumen Chakrabarti, and Walter Willinger. We next present a summary of the discussion that followed each question.

There has been talk of a new "network science". How does the scientific method apply to this new science? What are the steps a good network scientist should follow?

Much of Computer Science is about proving properties of models, not coming up with theories to explain scientific observations. Network science is new because validity is based on statistical evidence. In practical terms, there is a lot of funding for this research area and companies are taking advantage of large networks. It is exciting that people from many different backgrounds are working together to understand networks. From a different perspective, network science is not a new science. It is an old science created by sociologists. It is not clear what we are proving by experiments. However, the size of the networks makes a difference, allowing us to look at social networks under a new light. Furthermore, we have information networks, such as the Web.

Network science is not like natural science, because technology changes constantly. How does this change the game?

Technology is changing the landscape. Space is becoming smaller. Attention gets divided in different ways. Our behaviour and the ways we interact change, resulting in the need to change our assumptions. Our technology is creating a loop, in that we are creating objects and we study them as if we do not know what is inside, using the same tools as natural sciences, but applied to human-made artifacts. This type of research is going to earn the attribute "Science" for Computer Science because of this. The fact that mathematicians, physicists, and computer scientists are coming together is exciting, providing a good model of what should be happening more in academia. Economists also study complex systems, and they could join the effort.

Network science was invented by physicists, who are very good at popularizing their research. Physicists apply their approach to this field and make claims, then domain experts from computer science, engineering and mathematics come along and formalize the field, by uncovering the "truth" about these networks, and verifying experimentally their hypothesized properties. The process is messy, because the physical world does not change that fast, which is not true about the Web or online social networks, which change a lot faster. The commercial focus helps to ground the research, and keep it practical.

Will network science lead to the development of network engineering? What research directions do we need to pursue to get there?

Validation is really important. In much of this domain validation is not taken seriously. Validation will be different from one network to another. What is good validation is socially defined. Sometimes it takes a research community a long time to figure this out.

Should we develop benchmark datasets to support network science? What does it take to get them?

Benchmark datasets are necessary to promote the research. We have seen examples of this, for example TREC. We can even aim at building infrastructure to serve network science, for example the *Network Workbench* [6] from Indiana University, School of Library and Information Science. There is a danger with benchmark datasets of overfocusing the community on specific problems. We need a process of generating datasets with known patterns, for example KDD. Datasets are built for a specific purpose. Building a dataset is a lot of work. Perhaps a set of goals for network science must be formulated before we discuss benchmark datasets.

What are interesting goals for network science?

There are several tasks that can be identified. Clustering is one (of biological data for example). There is a variety of algorithms and implementations, and we need to define appropriate evaluation measures. Artificial datasets, where the boundaries between classes are controlled, may have an important role. Datasets have been used mostly for data fitting, which has the risk of overfitting a particular dataset. Finally, labelled datasets are needed, possibly by human judgment.

What claims can we make knowing the limitations of our network sampling techniques in terms of time scale and completeness? In other words, how do you figure out how sceptical to be when using data known to have defects?

Sampling a real network is a hard problem in general, and requires solid statistical expertise. Completeness is not achievable. Any sampling process may introduce bias. This is a fascinating area for statisticians, especially because outliers are meaningful. Hard problems involve prediction of very rare events. New methodologies are required, because of the datasets are massive and with many features. Web phenomena tend to have long tails. The question of the stability of observations needs to be considered. Special care is needed to establish the limitations of any analysis. The problem of estimation of the number of indexed pages by a search engine is addressed in [8]. Algorithm design must adapt to the changing nature of the data, for example, to account for the emergence of web spam.

Name your best open problem in this area

A variety of interesting problems was proposed: to come up with and test a model of how the network structure and the content co-evolve; to study the dynamics of graphs, including interaction between flows and connectivity; to involve social scientists in social network analysis; to analyze the reasons why a particular social networking service succeeds and to predict which of the many players will eventually take over; to study searching and ranking, given the diversity of indexed content on the Web; to address classical combinatorial problems related to large social networks.

---

[6] http://nwb.slis.indiana.edu/

3   Summaries of Poster Presentations

Neighborhood Watch: Document Classification in Typed Graphs

Ralitsa Angelova, Max-Planck Institute for Informatics

Classification is a challenging problem with a broad impact on areas like machine learning, information retrieval, pattern recognition, image analysis and bioinformatics. Recent research shows that incorporating relationships into the classification process is beneficial but poses difficulties which, if carelessly addressed, degrade the classification result. One hard problem is how to make use of the link structure in a heterogeneous environment. Such an environment is represented by a graph containing nodes of different types. Nodes of the same type as well as nodes that belong to different types are connected by edges (links). Different node types have different systems of possible class labels. The goal is to assign to each node the best suitable label among its possible classes according to a local likelihood and the node's neighborhood in the graph. A relaxation labeling approach is proposed for classification of heterogeneous graphs.

This is joint work with Prof. Gerhard Weikum.

Link Analysis Based Methods for Handling Abundance and Misrepresentation Over The Web

Amit Awekar, North Carolina State University

Broad aim of this work is to investigate how link analysis based methods can be useful to deal with abundance and misrepresentation issues over the Web. An algorithm SelHITS was introduced for answering broad-topic queries over the Web. The author aims to apply the same approach to other topic oriented tasks over the Web. Clustering hypertext repository is the current problem of interest. The proposed approach is to iteratively modify the representation of documents using link based ranking functions.

Towards Adaptive Web Search Engines

M. Barouni-Ebrahimi, University of New Brunswick

Web search engines efficiently surf the Internet and return the most relevant pages to the users' queries. However, the order of the recommended pages is not always in accordance with the users' priorities. The users needs to check the list of the recommended pages to find one of their interests. On the other hand, the queries sent by the users do not always corresponds to their intentions. The lack of user knowledge or unfamiliarity with the specific keywords and phrases in the domain knowledge leaves the user wondering about what phrases would be the most related ones to his desire. The contribution of this research is threefold. First, Complementary Phrase Recommender module suggests to the user a list of complementary phrases for his uncompleted query. Second, Related Phrase Advisor module provides a list of phrases related to the query segment that user has entered. These two modules guide the user to enter the more related phrases to his intention as a query. Third, Page Rank Revisor module refines the order of the recommended documents prepared by a conventional web search engine to help the user find the related web pages at top of the list.

This is joint work with Prof. Ali A. Ghorbani.

Evaluating Web Search Quality
Maxim Gurevich, Technion, Israel
Objectively assessing search quality is of great interest both to end-users and to search providers. Quality parameters like ranking quality, coverage of the web, index freshness, topic- and domain-specific coverage, and spam resilience are important for judging the effectiveness of search engines. Currently, search quality is evaluated mainly by manual techniques, using anecdotal test data. This makes the results difficult to reproduce and non-objective.

Random sampling is arguably the most efficient way to measure parameters on huge data sets, like search engines. Along sampling from the whole web/index of the search engine, sampling web pages from a given "topic" of the web/index (i.e., web pages in some domain, topic, language, etc.) may be interesting. Such samples can be used to measure the quality of search engines with respect to specific segments and domains. Evaluating ranking of search engines is another interesting area. Unlike current methods, which mainly rely on user studies, an automatically computable metric would be more statistically accurate and easy to reproduce. One idea is to use the latent human judgment in click-through data. The metric may then be used to compare rankings of major search engines.

Use of k-cores to characterize graph local structure
John Healy, Dalhousie University
A k-core of a graph is the subgraph generated by recursively removing all nodes with degree $< k$. This can be thought of as a weaker version of a clique. k-cores are useful for pruning low importance vertices. ncreasing the value k eliminates nodes resulting in a component either remaining in our graph, splitting into multiple components, or being eliminated entirely. The resulting small directed acyclic graph, capturing the evolution of components as k increases, reveals the structure of a graph. The the k-core representation is used in order to find which of several available generative models is the best description of a real world graph, by developing a method of summarizing the component trees for statistical comparison.

The Chromatic Number of Complex Networks
Paul K Horn, University of California, San Diego
The chromatic number of a graph, denoted $\chi(G)$ is a graph invariant which is deeply tied to the structure of the graph, as well as other important graph properties such as the independence number and clique number. The chromatic number of complex networks is considered by modeling complex networks as random graphs with given expected degree sequences through the $G(\mathbf{w})$ model introduced by Chung and Lu. A graph with a more general degree distribution $\mathbf{w} = (w_1, \ldots, w_n)$ is considered; letting $w = (w_1 + \ldots + w_n)/n$ denote the average expected degree. The work is based on a recent preprint of Frieze, Krivelevich and Smyth, in particular a condition guaranteeing that $\chi(G(\mathbf{w})) = \theta(w/\ln w)$; furthermore it was shown that if $\mathbf{w}$ fails this condition too badly, that indeed $\chi(G(\mathbf{w})) = \omega(w/\log w)$. An improved lower bound on $\chi(G(\mathbf{w}))$ was also given.

Some related questions were investigated. The proposed conditions on when $\chi(G(\mathbf{w})) = \omega(w/\log w)$ seem to suggest a better indicator of the chromatic number. Still open is the question of finding an asymptotic value of $\chi(G(\mathbf{w}))$. Deeply related to this question are questions regarding the size and number of independent sets in $G(\mathbf{w})$. A deeper understanding of these issues is likely necessary to asymptotically determine $\chi(G(\mathbf{w}))$ and is also interesting in its own right.

Graph theory in interconnection networks
Navid Imani, Simon Fraser University
The focus of the research lies somewhere between distributed computing and graph theory & enumerative combinatorics. The focus is on the problems arising in a wide variety of networks ranging from interconnection networks for multiprocessor systems and massively parallel systems to mobile and sensor networks and the WWW. Previous and on-going works by the author address well-known problems in such networks such as Load Balancing), Resource Placement, Intrusion detection, Distributed Data-Clustering, Combinatorial Properties of Existing and Newly Proposed Networks. Most papers involve theoretical modeling of the above mentioned issues and have deep roots in graph theory, enumerative combinatorics and algorithms. A current project involves network security where the probabilistic behavior of networks is studied in the face of different types of failures using a combination of approaches from probability theory and combinatorics.

How NAGA uncoils: Searching with Relations and Entities
Gjergji Kasneci, Max-Planck-Institut für Informatik
The everlasting enrichment of the Web with certain as well as uncertain and unstructured information calls for vertical search techniques which fulfill users' needs for querying the Web in a more precise way. Going one step beyond keyword search and allowing the specification of contextual concepts for keywords or relationships holding between them clears the way for new attractive and promising possibilities.

NAGA, a semantic search engine for the Web, was presented, which exploits relationships between entities for precise query specification and answering. NAGA's trump card is its ontological knowledge graph built on top of a refined data model which in turn serves as a basis for NAGA's query model and answer computation algorithms. NAGA extracts facts from Web pages and stores them into the above mentioned knowledge graph. Not only the extracted facts are recorded, but also a confidence measure for each fact is computed and maintained. NAGA provides a query language which can be capable of expressing queries ranging from simple keyword queries to complex graph queries which utilize regular expressions over relation names. NAGA's answer model is based on subgraph matching algorithms which in turn make use of intuitive scoring and ranking mechanisms. The approach we follow represents a general approach towards the semantic processing of information extracted from any unstructured text corpora.

This is joint work with Maya Ramanath, Fabian Suchaneck, and Gerhard Weikum.

Improving the Random-Surfer Model with Anonymized Traffic Data
Mark Meiss, Indiana University

Link-analytical algorithms for ranking Web search results such as Google's PageRank derive their power from the implicit statement of relevance made when the owner of one page decides to link to another. However, such methods are undermined by the fact that not all links are created equal: some are used much more often than others. The random-surfer model of PageRank assumes uniform distributions for starting locations, outgoing links to follow, and jump probabilities, but the behavior of actual surfers may be quite different. The aim of the present research is to gather large volumes of "click data" from anonymized packet captures of real HTTP sessions, analyze the extent to which this data does not reflect the random-surfer model, and develop a more sophisticated stochastic model in which the random distributions are based on the traffic patterns of actual users.

Neighborhoods in the Web Graph
Isheeta Nargis, Memorial University of Newfoundland

The World Wide Web can be represented by a large directed graph in which each vertex corresponds to a web page, and in which there is an arc from one vertex to another if there is a hyperlink between the corresponding web pages. It is infeasible to store and manipulate the entire Web Graph, so a focused approach is taken. Beginning with a specified web page, it is determined which other pages are in close proximity to it, and then the subgraph of the Web Graph is constructed that is induced by these pages (i.e. a Neighborhood Graph for the given initial vertex). Properties of these neighborhood graphs were investigated.

Communities in Large Networks: Identification and Ranking
Martin Olsen, University of Aarhus

The problem of identifying and ranking the members of a community in a very large network with link analysis only is studied, given a set of a (few) representatives of the community.

The concept of a *community* is defined justified by a formal analysis of a simple model of the evolution of a directed graph. It is shown that the problem of deciding whether non trivial communities exists is NP complete. Nevertheless experiments show that a very simple greedy approach can identify members of a community in the Danish part of the www graph with time complexity only dependent on the size of the found community and its immediate surroundings.

The members in a community are ranked by performing a computationally inexpensive calculation which is a "local" variant of the PageRank algorithm. The mathematical model behind the ranking is a small Markov Chain with the community as its state space forming a valuable basis for analyzing consequences of changes of the link structure.

Results are reported from a successful experiment on identifying and ranking Danish Computer Science sites.

Growing and classical protean graphs (new probabilistic models of the Web)

Pawel Pralat, Department of Mathematics and Statistics, Dalhousie University.

The Web may be viewed as a graph each of whose vertices corresponds to a static HTML web page, and each of whose edges corresponds to a hyperlink from one web page to another. Recently there has been considerable interest in using random graphs to model complex real-world networks to gain an insight into their properties.

An extended version of a new random model of the web graph is proposed in which the degree of a vertex depends on its age. The differential equation method is used to obtain basic results on the probability of edges being present. From this it is possible to characterize the degree sequence of the model and study its behaviour near the connectivity threshold.

The classical version of the model is also presented and the limit distribution of the 'recovery time' for connectivity near the connectivity threshold is characterized, and the diameter of the giant component.

This is a joint work with Tomasz Luczak and Nicholas Wormald.

Probabilistic models for concept discovery in unstructured text data

Mahdi Shafiei, Dalhousie University

Using probabilistic models for document and term clustering, document modeling and co-clustering has shown some major benefits over the traditional methods in recent years. In data mining research, these problems along with other problems including dimensionality reduction, topic segmentation, topic tracking and detection are closely related. These are also the fundamental building blocks of approaches to several applied problems including automatic summarization and machine translation. However, these problems have been approached independently of one another by the research community. The aim of the author is to bring all these interrelated problems under a single statistical model, and to exploit their interrelations. In previous work, hierarchical Bayesian models have been developed for clustering terms and documents. By the topic segmentation capability embedded in the model, the goal is to improve the clustering performance of the previous model on words and documents. Using the probabilistic Bayesian approach makes it possible to extend the proposed approach to a model capable of modeling topic tracking and shift in a principled way.

Shrack: A Pull-Only Peer-to-Peer Framework for Sharing and Tracking of Research Publications

Hathai Tanta-ngai, Dalhousie University

Shrack—a pull-only peer-to-peer framework for document sharing and tracking— is presented. Shrack is designed to support researchers in forming direct collaborations to autonomously share and keep track of new research publications based on their interests. A pull-only information dissemination protocol is used for peers to learn about document metadata of new research publications from peers having similar interests. A user's interest is represented by an automatically learned profile. Document metadata are viewed as semi-structured documents.

Peers first join the network using contacts acquired from real world collaboration, similar to exchanging email addresses or URLs. These contacts are used
as initial peer neighbours. Each peer can use the disseminated information to
build a local view of a semantic overlay network of peer interests, which represents groups of peers having similar semantic interests. Each peer can later use
the semantic overlay network to find new contacts of peers having a particular
interest, as well as search for documents archived by other peers. An overview
of the architecture of the system and research challenges were presented.

Privacy in Databases
Dilys Thomas, Stanford University

The explosive progress in networking, storage, and processor technologies has
resulted in an unprecedented volume of digital information. This has resulted
in an increased real-time processing of this digital information in streaming systems. In concert with this dramatic and escalating increase in digital data and
its real-time processing, concerns about privacy of personal information have
emerged globally. The ease at which data can be collected automatically, stored
in databases and queried efficiently over the internet has paradoxically worsened
the privacy situation, and has raised numerous ethical and legal concerns. These
concerns extend to the analytic tools applied to data. Problems arising from
private data falling into malicious hands include identity theft, stalking on the
web, spam etc. In the digital age, large amounts of confidential information are
accessible to hackers or insiders. Safeguards to protect the privacy of individuals, and security of society are becoming crucial for the effective functioning
of the Internet. Privacy enforcement today is being handled primarily through
legislation. The aim of this work is to provide technological solutions to achieve
a tradeoff between data privacy and data utility.

Web Mining putting emphasis on Web Graph Evolution monitoring
Akrivi Vlachou, Athens University of Economics and Business

The general focus of the research is on web mining and in particular on link
analysis and techniques for web graph representation. The web is a highly dynamic structure constantly changing. One of the biggest challenges is that of
searching the vast amounts of web graph data. The research area of web search
inherently involves the issue of page ranking. Research problems related to the
web graph evolution are addressed aiming at valid PageRank predictions and
monitoring the web-graph change. Additionally, a compact representation of the
web graph is envisioned, capitalizing on the changes of the web graph during
time. Such a representation will be used to effectively answer historical queries.

Statistical Analysis of Dynamic Communication Graphs
Xiaomeng Wan, Dalhousie University

Communication networks can be modeled as a dynamic graph with time-varying
edges. Real-life events cause communications that are unusual in either volume or
pattern in the graph. Given such a dynamic graph with embeded events, can we
detect when and where those events occur? The answer for this question is crucial for counter terrorism, network surveillance and traffic management. Most

event detection methods only focus on network-wide events. However, events associated with only a few individuals are more common and of significant interest, as well. In this project, a method is explored to detect those events with only local impacts. Three metrics to characterize communications from different viewpoints are proposed. Based on the variations of these metrics over time, local events are detected and characterized. Experiments on email data from our faculty show that these metrics are effective in identifying events, and the signals of the three metrics, combined in different ways, makes it possible to discriminate different types of events.

Dual Dynamic Programming and Reinforcement Learning
Tao Wang, University of Alberta

The dual approach to dynamic programming and reinforcement learning is investigated, based on maintaining an explicit representation of stationary distributions as opposed to value functions. A significant advantage of the dual approach is that it allows one to exploit well developed techniques for representing, approximating and estimating probability distributions, without running the risks associated with divergent value function estimation. A second advantage is that some distinct algorithms for the average reward and discounted reward case in the primal become unified under the dual. A modified dual of the standard linear program is presented that guarantees a globally normalized state visit distribution is obtained. With this reformulation, novel dual forms of dynamic programming are derived, including policy evaluation, policy iteration and value iteration. Moreover, dual formulations of temporal difference learning are derived to obtain new forms of Sarsa and Q-learning. Finally, these techniques are scaled up to large domains by introducing approximation, and develop new approximate off-policy learning algorithms that avoid the divergence problems associated with the primal approach. It is shown that the dual view yields a viable alternative to standard value function based techniques and opens new avenues for solving dynamic programming and reinforcement learning problems.

Sketching Landscapes of Page Farms
Bin Zhou, Simon Fraser University, Canada

The World Wide Web is a very large social network. It is interesting to analyze the general relations of web pages to their environment. For example, as rankings of pages have been well accepted as an important and reliable measure for the utility of web pages, it is worthwhile to understand generally how web pages collect their ranking scores from their neighbor pages.

Such information is not only interesting but also important for a few Web applications. (i) for Web spam, we can imagine identifying pages that receive a considerable amount of their ranking scores from bad pages; (ii) for Web page categorization, we could determine how much of page's ranking score comes from reputable pages from certain domains (e.g., the database and data mining community highly regards my page, but the network security community does not); (iii) for simple Web page characterization, it could be interesting to know that FedEx receives considerable link support from certain partner companies, etc.

In this research, the environment of web pages is modeled and its general distribution is analyzed. A novel web structure mining problem is studied, mining page farms, and its application is illustrated in link spamming detection. The general ideas and major contributions so far are as follows.

– A page farm is the set of pages contributing to (a major portion of) the PageRank score of a target page. The computational complexity of finding page farms is shown to be NP-hard. Then, a practically feasible greedy method is developed to extract approximate page farms.
– The statistics of landscapes of page farms are empirically analyzed using over 3 million web pages randomly sampled from the web. We have a few interesting findings.
– The application of page farms in spamming detection is investigated. Two spamicity measures are defined which can be used to detect spam pages, and evaluated on a real data set. The experimental results show that the methods are effective in detecting spamming pages.

References

1. Li, L., Alderson, D., Willinger, W., Doyle, J.: A first-principles approach to understanding the Internet's router-level topology. In: Proc. ACM SIGCOMM, pp. 3–14 (2004)
2. Doyle, J., Alderson, D., Li, L., Low, S., Roughan, M., Shalunov, S., Tanaka, R., Willinger, W.: The 'robust yet fragile' nature of the Internet. Proc. Nat. Acad. Sci. 102(41), 14497–14502 (2005)
3. Alderson, D., Chang, H., Roughan, M., Uhlig, S., Willinger, W.: The many facets of Internet topology and traffic. Networks and Heterogeneous Media 1(4), 569–600 (2006)
4. Chang, H., Jamin, S., Willinger, W.: To peer or not to peer: Modeling the evolution of the Internet's as-level topology. In: Proc. IEEE INFOCOM (2006)
5. Stutzbach, D., Rejaie, R., Duffield, N., Sen, S., Willinger, W.: On unbiased sampling for unstructured peer-to-peer networks. In: Proc. ACM/USENIX Internet Measurement Conference (IMC 2006) (2006)
6. Fabrikant, A., Koutsoupias, E., Papadimitriou, C.: Heuristically optimized tradeoffs: A new paradigm for power laws in the Internet. In: Widmayer, P., Triguero, F., Morales, R., Hennessy, M., Eidenbenz, S., Conejo, R. (eds.) ICALP 2002. LNCS, vol. 2380, pp. 110–122. Springer, Heidelberg (2002)
7. Menczer, F.: The evolution of document networks. Proc. Natl. Acad. Sci., USA, 101, 5261–5265 (2004)
8. Bar-Yossef, Z., Gurevich, M.: Random sampling from a search engine's index. In: WWW (2006)

Workshop on Algorithms and Models for the Web Graph

William Aiello[1], Andrei Broder[2], Jeannette Janssen[3], and Evangelos Milios[4]

[1] Department of Computer Science, University of British Columbia
[2] Yahoo! Research
[3] Department of Mathematics and Statistics, Dalhousie University
[4] Faculty of Computer Science, Dalhousie University

1 Overview

For barely a decade now the Web graph (the network formed by Web pages and their hyperlinks) has been the focus of scientific study. In that short a time, this study has made a significant impact on research in physics, computer science and mathematics. It has focussed the attention of the scientific community on all the different kinds of networks that have arisen through technology and human activity; some speak of a "new science of networks". It has brought the computational and deductive power of computer science to the study of the complex social networks formed by inter-human relationships. And, it has given birth to new branches of research in different areas of mathematics, most notably graph theory and probability.

The key event that focused attention on the link structure of the Web was the invention, by Brin and Page, of the PageRank algorithm [4]. PageRank is a method of ranking Web pages that is derived entirely from information about which page has hyperlinks to what other page, as opposed to what content of Web pages have. The phenomenal success of Google, the search engine built by Brin and Page, led to the realization that a lot of the information on the Web is contained in its link structure. Naturally, the *Web graph*, as this link structure came to be called, became the focus of scientific study.

Two experimental studies aimed to explore the Web graph. An extensive study by Broder et al. [5] analyzed a Web crawl obtained by the search engine *Altavista*. The second study, by Barabasi and Albert, used a more modest data set consisting of all URLs in the Web domain of Notre Dame University (`.nd.edu`). The studies reported similar findings. The most remarkable finding was that the distribution of the in-degrees (number of links pointing to a page) follows a *power law*. In particular the proportion of pages with degree k is proportional to $k^{-\gamma}$. A power law distribution has a heavy tail, which means that pages with high in-degree are relatively common.

Barabasi and Albert coined the term "scale-free network" for networks that exhibit a power law degree distribution. They proposed a growing graph model for such networks. The model is still one of the most widely used to explain the growth of self-organizing networks, and to simulate scale-fee networks. The leading principle of the model is that of *preferential attachment*: new nodes

W. Aiello et al. (Eds.): WAW 2006, LNCS 4936, pp. 18–23, 2008.

join the network, and link to existing nodes via a random process where the probability that an existing node receives a link is proportional to its current degree. In other words, nodes with higher degree have a greater chance of an increase in degree. It has been shown heuristically in [1], and more rigorously in [2] that this kind of reinforcement mechanism does produce a power law degree distribution. However, many aspects of this model have since been found to be not in accordance with the observed properties of real-life networks, as we will see in the review of Willinger's talk below. A variety of different models have been proposed, either based on variations on the preferential attachment theme, or on different principles such as copying, trade-off of contradictory objectives, and geometric embeddings of the nodes. We refer to [3] for an overview.

In this workshop, clearly the PageRank algorithm was still a central theme. The speakers presented various new angles to this theme. A general framework for ranking algorithms was presented by one of the invited speakers, and in this framework PageRank can be shown to be among the optimal ranking algorithms. Other speakers investigated the relation between PageRank and in-degree, both from an experimental and an analytical viewpoint. Another invited speaker showed how PageRank can be used as a tool in identifying graph communities. A third invited speaker investigated the question whether the use of PageRank to present search results leads to a reinforcement loop, where the ranking algorithm itself will lead to increased rank for already highly ranked pages.

Another central theme was graph modelling. One of the invited speakers, Fan Chung, is a central authority on graph modelling, and while her talk for WAW did not directly address graph modelling, she gave another talk in the associated winterschool where she presented a variety of important and challenging open questions on graph modelling. Another invited speaker questioned the premise of many of the early graph models, namely preferential attachment, and presented convincing arguments to persuade modellers to give greater weight to other aspects of the network, beyond the degree distribution. Other speakers analyzed various aspects of a variety of different random graph models, such as models obtained through a local optimization criterium that depends on an underlying geometry, and a model obtained through random perturbations of a pre-determined graph.

A variety of other themes was addressed by the contributed speakers and poster presenters. To name but a few: there were presentations on Web spam detection, network sampling, community identification, automatic Web-Site summarization, and Web mining based on both content and link structure. During the workshop, numerous informal discussions took place, and quite a few led to new collaborations and ideas. Also, the workshop inspired a number of participants to initiate the organization of WAW 2007, which will take place in San Diego.

2   Ranking of Search Results

With the enormous quantity of information now available on the Web, it is clear that information retrieval is only a minor part of the task of a search engine.

Most important is the *ranking* of the retrieved results. In particular, since most users do not browse beyond the first page of search results, the selection of the top ten of highest rank pages is crucial. The link-based PageRank algorithm has proven its worth through the Google search engine. Google's ranking method is top-secret, but it is clear that PageRank is still central to its success.

The PageRank algorithm ranks Web pages according to their *PageRank value*. The PageRank value of page i is a number between 0 and 1 denoted by $PR(i)$. The original description of PageRank presented in [4] is as follows: PR is a vector indexed by all the nodes in a graph that satisfies

$$PR(i) = c \sum_{j \to i} \frac{1}{d_j} PR(j) + (1 - c),$$

where d_j is the out-degree (number of outgoing links) of node j, and c is a constant between 0 and 1, often called the *teleportation factor*. The vector PR is the stationary distribution of a Markov chain representing a random surfer: at each step, the surfer either follows one of the links of the current page with equal probability, or "teleports" to any existing page with equal probability. The probability with which the first option is chosen is given by c. If this process is ergodic the PageRank vector is the principal eigenvector of the transition matrix of this process, the socalled *PageRank matrix*.

Since its original introduction, many aspects and variations of the PageRank algorithm have been studied (for an overview, see [6]. Also, a number of other link-based ranking algorithms have been proposed. We note the HITS algorithm by Jon Kleinberg, which was developed around the same time as PageRank. The HITS algorithm assigns two values to each node, a hub value and an authority value. If A is the incidence matrix of the graph (so that $A_{ij} = 1$ iff (i, j) is an edge), then the vector of hub values is the principal eigenvector of AA^T, and the vector of authority values is the principal eigenvector of $A^T A$. Hub- and authority values can be calculated iteratively, similar to the PageRank computation. Another ranking algorithm is SALSA, a stochastic algorithm that combines ideas from PageRank and HITS.

The question whether PageRank has a self-reinforcing effect was discussed in an invited talk titled *Googlearchy of Googlocracy? How search affects Web traffic and growth* by Filippo Menczer of Indiana University. Menczer explained that search engines bias the Web traffic through their ranking strategy. Namely, it is known that most Web users will rely on a search engine to navigate the Web, and are likely to view only the first page of results returned by the search engine. Hence, they will only visit the handful of pages that are most highly ranked.

Since Google is by far the most popular search engine, and is known to rely on the PageRank algorithm for its ranking, some have argued that this creates a vicious cycle where pages with a high PageRank are more visible to creators of new pages, and thus more likely to get linked to. The increased number of in-links, in turn, increases the page's PageRank score, thus leading to a "rich-get-richer" situation where a small number of pages will dominate the high ranks of the search engine.

Menczer convincingly showed that, contrary to these claims and his own intuition, the use of search engines actually has an egalitarian effect. Empirical evidence and theoretical analysis, carried out by Menczer in collaboration with Santo Fortunato, Alessandro Flammini, and Alessandro Vespignani, showed that, in fact, search engines mitigate the attraction of popular pages, and direct more traffic toward less popular sites.

The PageRank algorithm also was the topic of a number of contributed talks. Specifically, talks by Nelly Litvak (Twente) and Alessandro Flammini (Indiana) examined the relation between PageRank and in-degree from different angles. Litvak considered the question from a theoretical viewpoint, employing sophisticated tools from probability theory. Flammini's work combined experimentation with a more heuristic mean field analysis.

Fan Chung Graham (UCSD), another invited speaker, described how PageRank can be used to partition a graph. Graph partitioning is an important tool for identifying Web communities. In turn, identifying Web communities helps fine tune search results, and can increase our understanding of the Web. Chung showed how the ordering of the nodes produced by the PageRank vector reveals a location where the graph can be cut with minimal loss of "flow" through the graph.

Finally Soumen Chakrabarti, of the Indian Institute of Technology in Mumbai (IITB), considered in his invited talk the problem of finding an appropriate ranking in the more general setting of entity-relation (E-R) graphs. The general framework is still that of nodes connected by links, and text associated with each node, but now both nodes and links can be annotated, assigned to categories, etc. However, the extractors and annotators may be imperfect and incomplete. In this sense, E-R graphs are a representation of semi-structured text. This representation makes ranking difficult: nodes and edges have diverse semantics and are not equally important. Their importance may even vary by query. Consequently, there is no single successful ranking function for general E-R graph search applications. On the other hand, the information is too complex for completely manual tuning. Chakrabarti explained how machine learning techniques can be employed to automatically tune the ranking function to the data and query at hand.

3   Models for the Web Graph and Other Complex Networks

A second theme of the workshop was that of stochastic models for the Web graph in particular, and self-organizing networks in general. A self-organizing network is an evolving network, formed in a de-centralized manner by a number of individual agents. Each agent determines its own link environment, usually with only limited knowledge of the entire network. The World Wide Web is a prime example of a self-organizing network: the links to other Web pages that a given page contains are determined by the "agent" creating the page, not by any central authority. Other examples of self-organizing networks are: networks formed by scientific papers and their citations, law cases and their references, social networks (both of humans and animals), and networks modelling the interaction of proteins in a cell (protein-protein interaction, or PPI, networks).

As mentioned in the introduction, the first generation of models was mainly based on the principle of *preferential attachment*. In models based on preferential attachment, new nodes joining the network choose a pre-determined number of neighbours, chosen with probability proportional to the degree. In other words, nodes with high degree have a better chance of receiving a link to the new node. The first use of preferential attachment as a method to generate graphs with a power law was first proposed by [1]. More advanced models based on the principle can be found in [7,8].

The invited talk by Walter Willinger, of AT&T Research Labs, sharply questioned the use of the preferential attachment principle as a catch-all to explain power law graphs. In fact, Willinger started out with a criticism of the various studies that measured power law degree distributions in various real-life networks, especially those associated with the internet. Namely, accurate measurements of connectivity-related parameters of the Internet are notoriously hard to obtain. Willinger then argued that the evolution of the internet is more likely driven by restrictions that arise from the technological constraints of various components that determine the physical internet. He proposed an alternative approach to modelling, that relies heavily on domain knowledge. This approach is capable of explaining a wide range of diffeerent system behaviours and provides a basis for exploring when and when not to expect a power law degree distribution.

The second part of the invited talk by Filippo Menczer also addressed the question of modelling. After discussing the mutual interaction of the ranking of search results by search engines and Web user traffic, as described in the previous section, Menczer introduced a graph generation model based on the principle of search-driven network growth. This model is based on a principle of preferential attachment based on a ranking of the nodes. In a moment of synergy, it turned out that one form of this model was a special case of a graph model presented in a poster by Pawel Pralat. Discussions between Menczer and Pralat have since led to new research combining both ideas.

Several contributed papers and posters also had graph models as their topic. Abraham Flaxman presented a talk on how the addition of a random perturbation to a given graph influences the expansion properties. This work can lead to tools to analyze sensitivity of traffic flow to small changes in network topology. Ross Richardson investigated a geometric graph model based on an optimization criterium which is a trade-off between optimization of global and local connections. Anthony Bonato gave a talk on the infinite limit graphs that arise when time is going to infinity in graph generation models, and on what these infinite limits can tell us about the model. Posters by Gao and Pralat presented new models, while a poster by Healy gave a new method to evaluate and compare different models. The posters by Healy and Pralat were awarded prizes in a poster contest where all WAW participants were judges.

4   Overview of the Workshop

This workshop falls into a series of workshops on the Web graph: WAW '02 in Vancouver, WAW '03 in Hungary, and WAW '04 in Rome. The series was started by Andrei Broder as a forum for the latest research developments related to the Web graph in particular, and other complex networks in general. It is important that such a forum exists, because the interdisciplinary nature of this type of research has the unfortunate effect that it is often hard to find the right audience for presenting the results.

Researchers in the emerging area of algorithms and models for the Web graph are scattered geographically, and belong to disciplines with quite different academic cultures. The excellent research environment at BIRS was ideally suited to foster an understanding between people from different backgrounds. Discussions over lunch and dinner led to interesting synergies. The informal atmosphere fostered by the BIRS facilities fostered a climate that favoured an active audience. Talks were often interrupted by questions, which then led to discussions that were continued during the breaks.

Young researchers had a chance to showcase their research during a poster session. All participants in the workshop actively participated in the poster session, and the keynote speakers and other senior researchers naturally assumed a mentoring role and offered comments and suggestions.

In all, we believe the workshop highly successful, and we are grateful to BIRS, MITACS, Genieknows.com and Yahoo Inc. for their help in making it happen.

References

1. Barabási, A., Albert, R.: Emergence of scaling in random networks. Science 28, 509–512 (1999)
2. Bollobás, B., Riordan, O., Spencer, J., Tusnády, G.: The degree sequence of a scale-free random graph process. Random Structures and Algorithms 18, 279–290 (2000)
3. Bonato, A.: A survey of models of the web graph. In: Hamel, López-Ortiz (eds.) CAAN 2006. LNCS, Springer, Heidelberg (2004)
4. Brin, S., Page, L.: The Anatomy of a Large-Scale Hypertextual Web Search Engine. Computer Networks and ISDN Systems 33, 107–117 (1998)
5. Broder, A., Kumar, R., Maghoul, F., Rahaghavan, P., Rajagopalan, S., State, R., Tomkins, A., Wiener, J.: Graph structure in the web. In: Proceedings of the 9th International World-Wide Web Conference (WWW), pp. 309–320 (2000)
6. Langville, A., Meyer, C.: Deeper inside PageRank. Internet Mathematics 1(3), 335–380 (2004)
7. Aiello, W., Chung, F., Lu, L.: Random evolution in massive graphs. In: Abello, J., et al. (eds.) Handbook on Massive Data Sets, pp. 97–122. Kluwer Academic Publishers, Dordrecht (2002)
8. Cooper, C., Frieze, A.: On a general model of web graphs. Random Structures and Algorithms 22, 311–335 (2003)

Expansion and Lack Thereof in Randomly Perturbed Graphs

Abraham D. Flaxman

Microsoft Research
Redmond, WA, USA, 98052
abie@microsoft.com

Abstract. This paper studies the expansion properties of randomly perturbed graphs. These graphs are formed by, for example, adding a random 1-out or very sparse Erdős-Rényi graph to an arbitrary connected graph.

The central results show that there exists a constant δ such that when any connected n-vertex base graph $\bar{G}$ is perturbed by adding a random 1-out then, with high probability, the resulting graph has $e(S, \bar{S}) \geq \delta|S|$ for all $S \subseteq V$ with $|S| \leq \frac{3}{4}n$. When $\bar{G}$ is perturbed by adding a random Erdős-Rényi graph, $\mathbb{G}_{n,\epsilon/n}$, the expansion of the perturbed graph depends on the structure of the base graph. A necessary condition for the base graph is given under which the resulting graph is an expander with high probability.

The proof techniques are also applied to study rapid mixing in the small worlds graphs described by Watts and Strogatz in [*Nature 292* (1998), 440–442] and by Kleinberg in [*Proc. of 32nd ACM Symposium on Theory of Computing* (2000), 163–170]. Analysis of Kleinberg's model shows that the graph stops being an expander exactly at the point where a decentralized algorithm is effective in constructing a short path.

The proofs of expansion rely on a way of summing over subsets of vertices which allows an argument based on the First Moment Method to succeed.

1 Introduction

Developing models of complex networks has been a major industry in the fields of physics, mathematics, and computer science during the last decade. Empirical studies of many large networks gleaned from the real world have revealed that, unlike the classical models of Erdős-Rényi random graphs developed for applications to probabilistic combinatorics, many of the complex networks which surround us today have high clustering coefficients and power-law degree distributions. This observation has driven the development of numerous alternative distributions for random graphs, which often are described by some generative procedure.

Unfortunately, it is much easier to propose a generative procedure than to refute one, which has led to the preponderance of models available today. However, the copious models of real-world graphs may not withstand the test of time

W. Aiello et al. (Eds.): WAW 2006, LNCS 4936, pp. 24–35, 2008.

any better than the Erdős-Rényi distribution. This motivates the approach pursued in the present paper. Instead of studying a particular model for generating graphs with the hopes of finding it "more realistic" than previously proposed models, this paper considers an approach for incorporating randomness into network modeling that is less model-specific.

In this paper, a complex network is viewed as composed of a *base graph* and a *random perturbation*. The general goal in this framework is to show that some property is likely to hold for a wide variety of base graphs and under a very gentle random perturbation. For example, [1] shows that if a network is generated from any connected base graph on n vertices, perturbed by taking the symmetric difference with ϵn random edges, then, **whp**[1], if the network is connected then it will have diameter $\mathcal{O}(\epsilon^{-1} \log n)$.

This directly extends Bollobás and Chung's pioneering study of a cycle plus a random matching [2], and can be viewed as work in the line of "How many random edges make a dense graph Hamiltonian?", and subsequent studies of the effects of adding a few random edges to dense graphs [3,4,5]. It is also similar to the smoothed analysis of algorithms introduced by Spielman and Teng in [6], which has been used to explain why algorithms perform better in practice than worst-case bounds predict. Also similar are the hybrid graphs studied in [7] which explicitly model long and short edges.

In addition to the perturbation models like those considered on sparse random instances in [1], this paper will consider non-uniform perturbations, in the spirit of Jon Kleinberg's small-world model [8] and long-range percolation in finite graphs studied in [9,10,11], and also the graph which both these models build upon, the small-world model of Watts and Strogatz [12].

1.1 Results and Applications

The main technical development in this paper is a technique for understanding when randomly perturbed graphs exhibit expansion properties. This is motivated by the success of expansion bounds on more traditional random instances. For sufficiently dense Erdős-Rényi graphs, the First Moment Method provides a simple way to obtain a **whp** lower-bound on expansion. This paper provides a new method of accounting that permits a similar First-Moment-Method approach to be employed on randomly perturbed graphs.

For clear presentation, this new application of the First Moment Method is presented in the proof of an expansion property for a random graph G formed by perturbing any connected graph $\bar{G}$ by adding a random 1-out (which is the random graph formed by adding an edge from each vertex to another vertex chosen uniformly at random, and then ignoring the directions of the edges).

Theorem 1. *For any sufficiently small $\delta > 0$, for any n-vertex connected graph $\bar{G}$, and for $R \sim \mathbb{G}_{n,1\text{-}out}$, the perturbed graph $G = \bar{G} + R$ has the following property* **whp**: *for all $S \subset V$ with $|S| \leq \frac{3}{4}n$, at least $\delta|S|$ edges go between S and $\bar{S}$.*

---

[1] In this paper *with high probability* (**whp**) means that a sequence of events $\{\mathcal{E}_n\}$ has $\Pr[\mathcal{E}_n] \to 1$ as $n \to \infty$.

Table 1. Conditions for expansion under several perturbations

Perturbation	Expander **whp**?
1-out	Yes, for any connected G
$\mathbb{G}_{n,\epsilon/n}$	Not if $\bar{G}$ has a bad partition
Watts-Strogatz Small World	Not if $\bar{G}$ has a bad partition
Kleinberg Small World	Yes, for any conn. $\bar{G}$, if $r < r_{\max}(\bar{G})$

This technology is also applied to similar random graphs, to yield results summarized in Table 1.

1.2 History of Expansion in Random Graphs

A close connection between edge expansion, vertex expansion, spectral gap, and mixing time has emerged over the last 40 years [13,14,15,16,17]. Through this link, many different results on random graphs can be related to expansion properties. In regular and nearly-regular random graphs, bounds on the second-largest eigenvalue of the adjacency matrix give bounds on expansion, [18,19,20,21,22]. In a graph with a power-law degree distribution or other far-from-regular graphs, the eigenvalues of the adjacency matrix are not necessarily related to eigenvalues of the Laplacian and expansion. Both have been investigated theoretically and experimentally in recent years [23,24,9,25,26,27].

In the empirical study of complex networks occurring in the real world, examining Laplacian eigenvalues has revealed that some real networks are expanders and others are not [28,29]. This has led to the development of web graph models which specifically avoid being good expanders [30].

Algorithmically, there are many benefits to knowing that a graph is an expander (for example, rapid mixing, disjoint paths and routing, and robustness to attacks) and there are many other benefits to knowing that a graph is not an expander (for example, high-quality cuts, divide-and-conquer algorithms, and compressing data). Expansion may be less universal to real-world graphs than other properties observed empirically like local clustering and power-law degree distributions.

1.3 Notation

Undirected edges are sets of 2 vertices, but edge $\{u,v\}$ will be abbreviated as uv when it is not confusing to do so. For any graph H, let $E(H)$ denote the edge set of H, let $V(H)$ denote the vertex set of H, and for sets $S,T \subseteq V(H)$, let $e_H(S,T)$ denote the number of edges between S and T in H, and let $e_H(S)$ denote the number of edges in the graph induced by vertex set S (the induced graph is denoted $H[S]$). Let $\deg_H(v)$ denote the degree of v in H. The subscripts for $e(S,T)$, $e(S)$, and $\deg(v)$ will be omitted when referring the graph G if it is not too confusing to do so.

1.4 Distributions for Random Graphs

Perturbed graph 1 ($\mathcal{P}_1$): The randomly perturbed graph that appears in Theorem 1 is a random graph generated by starting with base graph $\bar{G}$ and adding a random 1-out ($\mathbb{G}_{n,1\text{-out}}$ is the distribution of random graphs where every vertex chooses a neighbor uniformly at random and adds an edge to it.) The random graph $G = \bar{G} + R$ where $R \sim \mathbb{G}_{n,1\text{-out}}$ is studied primarily to illustrate the central technique of this paper, although it is a reasonably small perturbation. On average it changes the degree of every vertex by 2.

Perturbed graph 2 ($\mathcal{P}_2$): In the context of studying the effects of randomness in complex networks without making drastic assumptions about the distribution of the randomness, it would be better to use a perturbation that does not change the base graph as much as a 1-out does. This can be accomplished by starting with base graph $\bar{G}$ and adding a sparse Erdős-Rényi random graph ($\mathbb{G}_{n,\epsilon/n}$ is the distribution of random graphs where each of the $\binom{n}{2}$ candidate edges appears independently with probability ϵ/n.) The random graph $G = \bar{G} + R$ where $R \sim \mathbb{G}_{n,\epsilon/n}$ is studied in [1], which shows that **whp** $\operatorname{diam}(G) = \mathcal{O}(\epsilon^{-1} \log n)$. Since, on average, this perturbation changes the degree of every vertex by only ϵ, the local effects of the perturbation are quite minimal.

Small-world graph 1 ($\mathcal{SW}_1$): The small-world model of Watts and Strogatz is generated by starting with a base graph $\bar{G}$ and an ordering of the edges $E(\bar{G})$ (in [12], $\bar{G}$ is a ring of n vertices with each vertex connected to its k nearest neighbors with $k \gg \ln n$, and the edges are ordered in a particular way that is implicit in the description of the perturbation). The base graph is perturbed in the following fashion: proceed through the edges according to the ordering, and for each edge, with probability p, randomly rewire this edge to a vertex chosen uniformly at random, with duplicate edges forbidden; otherwise leave the edge in place.

Small-world graph 2 ($\mathcal{SW}_2$): Kleinberg's small-world graph is a random digraph generated by starting with a base graph $\bar{G}$ and a distance function $d(\cdot, \cdot)$ on the vertices of $V(\bar{G})$ (in [8], $\bar{G}$ is primarily taken to be an $n \times n$ grid, where $V = [n]^2$, and uv is an edge if $d_1(u, v) \leq p$; the distance function is taken to be the ℓ_1 norm). The base graph is perturbed by adding q random edges out of every vertex independently at random, where the i-th edge out of vertex v is denoted by $e_{v,i}$ and is chosen according to the distribution $\Pr[e_{v,i} = vw] = d(v, w)^{-r} / \left(\sum_{u \neq v} d(v, u)^{-r} \right)$ for all $w \neq v$ (here r is a parameter of the model).

Comparison of $\mathcal{SW}_1$ and $\mathcal{SW}_2$: $\mathcal{SW}_2$ is often viewed as a generalization of $\mathcal{SW}_1$. The big difference is that, while $\mathcal{SW}_1$ rewires edges uniformly at random, $\mathcal{SW}_2$ includes the parameter r, which controls the degree to which the underlying network is willing to try new things.

There is also a subtle difference between these two models. While $\mathcal{SW}_1$ randomly rewires each edge of the underlying graph with probability p (which, for a d-regular graph, results in dp random edges expected out of each vertex), $\mathcal{SW}_2$

adds q random edges out of each vertex. This sounds very similar for $q = dp$, and it is similar, but it is also different, in a very important way. Graphs from the $\mathcal{SW}_2$ distribution are expanders **whp**, while graphs from the $\mathcal{SW}_1$ distribution are not necessarily so.

1.5 Outline of What Follows

Section 2 proceeds with the proof of Theorem 1, which uses a new method of First-Moment-Method accounting to show that $G = \bar{G} + R$ has $e(S, \bar{S}) \geq \delta|S|$ for all S **whp** when R is a 1-out ($\mathcal{P}_1$).

Section 3 considers the more gentle perturbation, where R is distributed as $\mathbb{G}_{n,\epsilon/n}$ instead of as a 1-out. In this case, G is not necessarily an expander, and a criteria for $\bar{G}$ of having a "bad partition" is shown to prevent G from satisfying the expansion property **whp**. The same results are also shown to hold for Watts-Strogatz random graphs ($\mathcal{SW}_1$). In particular, when $\bar{G}$ is a cycle with edges connecting each vertex to its k nearest neighbors, or when $\bar{G}$ is a d-dimensional grid, it contains a bad partition and hence the perturbed graph is not an expander **whp**.

Section 4 considers the $\mathcal{SW}_2$ perturbation, where $\bar{G}$ is perturbed by a non-uniform q-out, in which each random edge out of v chooses a vertex w with probability related to distance from v to w under some distance function $d(\cdot, \cdot)$, according to $\Pr[e_{v,i} = vu] = d(v,u)^{-r}/\sum_{w \neq v} d(v,w)^{-r}$. For $q = 1$ and $r = 0$, this reduces to the base-graph-plus-1-out considered in Section 2, and the techniques from that section are shown to extend for $r > 0$ for grid-like graphs. These techniques show that when $\bar{G}$ is a d-dimensional grid, G is an expander for any $r < d$, and furthermore, there is a threshold at $r = d$, at which point G is no longer an expander.

This shows that the transition from expanding to non-expanding occurs precisely at the point where a local algorithm can find polylogarithmic length paths in the network.

2 Perturbing Any Connected $\bar{G}$ with a 1-Out Yields Expander

The proof of Theorem 1 is an application of the First Moment Method, and relies on a moderately precise calculation of the expected number of sets S which violate the bound $e(S, \bar{S}) \leq \delta|S|$. This is achieved by considering separately the sets with $|S| \leq \gamma n$ and $|S| > \gamma n$ for an appropriately chosen constant γ.

Proof of Theorem 1: A straightforward way to obtain an upper bound on the probability that there exists a set $S \subseteq V$ with $|S| \leq \frac{3}{4}n$ and $e(S, \bar{S}) \leq \delta|S|$ is the following: let Z_S be an indicator random variable for the event that a particular set S satisfies these conditions, and calculate an upper bound on the expected value of the sum $Z = \sum_{S \subseteq V} Z_S$. Showing that the expected value tends to 0 as n tends to ∞ yields a bound which proves the theorem, because for any non-negative random variable, $\Pr[Z \geq 1] \leq \mathrm{E}[Z]$ (this deceptively simple inequality is honored with the title "The First Moment Method"; see, for example, [31]).

Unlike the simple application of the First Moment Method, which is sufficient to show that $G \sim \mathbb{G}_{n,k\text{-out}}$ is likely to be an expander when k is a large enough constant, to make this calculation yield results about a perturbed graph will require considering the structure of the set S in the base graph $\bar{G}$.

The key trick is to use a special tour of $\bar{G}$ to describe each set S; let T be a spanning tree of $\bar{G}$, and let $W = (e_1, e_2, \ldots, e_{2(n-1)})$ be an Euler tour of the multigraph formed by including every edge of T twice. That is to say that W is a sequence of edges which gives a circuit in G that traverses each edge of T exactly 2 times. Such a tour exists because doubling every edge of T makes the degree of every vertex even. For any set S, let $\mathbf{I}_S \in \{0,1\}^{2(n-1)}$ be the incidence vector with $\mathbf{I}_S(i) = 1$ iff $e_i \in E(T[S])$. Let $e(\mathbf{I}_S) = |\{i : \mathbf{I}_S(i) \neq \mathbf{I}_S(i+1)\}|$ denote the number of times the Euler tour crosses the boundary of S. There is a direct relationship between $e_T(S, \bar{S})$ and $e(\mathbf{I}_S)$. Since each edge of T appears twice in W,

$$e_{\bar{G}}(S, \bar{S}) \geq e_T(S, \bar{S}) \geq e(\mathbf{I}_S)/2. \tag{1}$$

To obtain a bound on the expected value of the sum $\sum_{S:|S|=s} Z_S$, let

$$\mathcal{S}_{s,k} = \{S : |S| = s, e(\mathbf{I}_S) = k\}$$

denote the collection of sets S of size s for which T crosses the boundary of S exactly k times. Since every S maps to a unique $\mathbf{I}_S$, it follows that

$$|\mathcal{S}_{s,k}| \leq 2\binom{2n}{k},$$

because an incidence vector with k changes in value can be described by giving the k "change positions" and specifying if the first bit is a 0 or a 1.

For $S \in \mathcal{S}_{s,k}$, equation (1) shows that $e_{\bar{G}}(S, \bar{S}) \geq k/2$, so, in order to have $e(S, \bar{S}) \leq \delta s$, it is necessary that $e_{\bar{G}}(S, \bar{S}) \leq \delta s$ and $e_{R \setminus \bar{G}}(S, \bar{S}) \leq \delta s - k/2$. This implies that $e_R(S, \bar{S}) \leq 2\delta s - k/2$, which is impossible when $k > 4\delta s$. Thus,

$$\sum_{S:\,|S|=s} \mathrm{E}\,[Z_S] \leq \sum_{k=1}^{4\delta s} \sum_{S \in \mathcal{S}_{s,k}} \Pr\left[e_R(S, \bar{S}) \leq 2\delta s\right]$$

$$\leq \sum_{k=1}^{4\delta s} 2\binom{2n}{k} \Pr\left[e_R(S, \bar{S}) \leq 2\delta s\right]$$

$$\leq (8\delta s)\binom{2n}{4\delta s} \Pr\left[e_R(S, \bar{S}) \leq 2\delta s\right]$$

$$\leq n\left(\frac{ne}{2\delta s}\right)^{4\delta s} \Pr\left[e_R(S, \bar{S}) \leq 2\delta s\right], \qquad \text{for } \delta \leq 1/8.$$

Finishing the calculation requires an upper-bound on $\Pr\left[e_R(S, \bar{S}) \leq 2\delta s\right]$, for which it is necessary to consider separately the large and small sets S.

Large sets expand: When $s = |S| \geq \gamma n$, $\mathrm{E}[e_R(S, \bar{S})] \geq s\left(1 - \frac{s}{n}\right)$ and Chernoff's bound (see, for example, [32]) gives

$$\Pr\left[e_R(S, \bar{S}) \leq 2\delta s\right] \leq \exp\left\{-s\left(1 - \frac{s}{n}\right)\left(1 - \frac{2\delta}{1 - s/n}\right)^2 / 2\right\}.$$

So, for $\gamma n \leq s \leq \frac{3}{4}n$,

$$\sum_{S:|S|=s} \mathrm{E}[Z_S] \leq n\left[\left(\frac{e}{2\delta\gamma}\right)^{4\delta} \exp\left\{-\frac{(1 - 8\delta)^2}{8}\right\}\right]^s.$$

For any constant γ, if δ is a sufficiently small constant then this upper-bound is exponentially small in n.

Small sets expand: When $s = |S| \leq \gamma n$, a tighter bound on the probability can be obtained directly by

$$\Pr\left[e_R(S, \bar{S}) \leq 2\delta s\right] \leq \binom{s}{2\delta s}\left(\frac{s}{n}\right)^{s - 2\delta s} \leq \left[\left(\frac{e}{2\delta}\right)^{2\delta}\left(\frac{s}{n}\right)^{1 - 2\delta}\right]^s.$$

So

$$\sum_{S:|S|=s} \mathrm{E}[Z_S] \leq n\left[\left(\frac{ne}{2\delta s}\right)^{4\delta}\left(\frac{e}{2\delta}\right)^{2\delta}\left(\frac{s}{n}\right)^{1 - 2\delta}\right]^s = n\left[\left(\frac{e}{2\delta}\right)^{6\delta}\left(\frac{s}{n}\right)^{1 - 6\delta}\right]^s.$$

For $\frac{3}{1 - 6\delta} \leq s \leq \gamma n$ and δ sufficiently small, this upper-bound is $o(1/n)$.

Tiny sets expand: For $\delta \leq \frac{1}{12}$, the tiny sets S, of size $s \leq \frac{3}{1 - 6\delta}$, will satisfy $e(S, \bar{S}) \geq \delta s$ because the base graph $\bar{G}$ is connected and so $e(S, \bar{S}) \geq 1 \geq \delta\frac{3}{1 - 6\delta}$.
Putting this all together shows that there exists $\delta > 0$ such that

$$\Pr\left[\text{exists } S : |S| \leq \frac{3}{4}n \text{ and } e(S, \bar{S}) \leq \delta|S|\right] \leq \sum_{s=1}^{\frac{3}{4}n} \sum_{S:|S|=s} \mathrm{E}[Z_S] = o(1).$$

$\square$

3   Gentler Perturbation Does Not Necessarily Yield Expander

Adding a 1-out to a graph increases the average degree of a vertex by 2. This is not much, but it is not nothing. This section investigates the effects of perturbing $\bar{G}$ by adding a random instance of $\mathbb{G}_{n, \epsilon/n}$ (which is the Erdős-Rényi graph where every candidate edge is included independently with probability ϵ/n). The intention of the parameterization ϵ/n is to indicate that ϵ should be thought of as a small constant, although the results of this section apply to any constant ϵ.

An attempt to show that if $R \sim \mathbb{G}_{n,\epsilon/n}$ then the perturbed graph $G = \bar{G} + R$ is an expander can begin by following in the footsteps of the proof of Theorem 1. And such a proof attempt will succeed in showing that **whp** the large sets in G expand.

Theorem 2. *For any $\epsilon > 0$, for any sufficiently small $\delta > 0$, for any n-vertex connected graph $\bar{G}$, and for $R \sim \mathbb{G}_{n,\epsilon/n}$, the perturbed graph $G = \bar{G} + R$ has the following property* **whp**: *for all $S \subseteq V$ with $e^{-\epsilon/(64\delta)}n \le |S| \le \frac{3}{4}n$, at least $\delta|S|$ edges go between S and $\bar{S}$.*

Proof. Follow the proof of Theorem 1. The only new calculation this proof requires is a fresh application of Chernoff's bound. For this perturbation, $E[e_R(S, \bar{S})] = \epsilon s \left(1 - \frac{s}{n}\right)$, and so

$$\Pr[e_R(S, \bar{S}) \le \delta s] \le \exp\left\{ -\epsilon s \left(1 - \frac{s}{n}\right) \left(1 - \frac{\delta}{\epsilon(1 - s/n)}\right)^2 \Big/ 2 \right\}.$$

$\square$

However, following the proof of Theorem 1 does not succeed in showing that small sets expand. And indeed, it should not show this, because it is not necessarily true. If $\bar{G}$ has a *bad partition* (defined below) then **whp** G is not an expander.

Definition 3. *$\bar{G}$ has a δ-bad partition iff $V(\bar{G})$ can be partitioned into sets $S_1, \ldots, S_k, \bar{S}$ for which the following inequalities hold:*

$$|S_i| \le \frac{1}{2}\epsilon^{-1} \ln n, \qquad \qquad \text{for } i = 1, \ldots, k;$$
$$e_{\bar{G}}(S_i, \bar{S}_i) < \delta|S_i|, \qquad \qquad \text{for } i = 1, \ldots, k;$$
$$k = \omega(n^{1/2}).$$

Theorem 4. *For any $\epsilon > 0$, any $\bar{G}$, and $R \sim \mathbb{G}_{n,\epsilon/n}$, if $\bar{G}$ has a δ-bad partition, then* **whp** *there exists $i \in \{1, \ldots, k\}$ such that $e_R(S_i, \bar{S}_i) = 0$, and hence $G = \bar{G} + R$ has $e(S_i, \bar{S}_i) < \delta|S_i|$.*

The proof is a direct application of the Second Moment Method (as described in [31]) and omitted due to lack of space.

This theorem applies to show that, for example, when $\bar{G}$ is the d-dimensional grid graph, the perturbed graph is not an expander **whp**.

Corollary 5. *Let $\bar{G}$ be a d-dimensional grid on $N = n^d$ vertices and let $R \sim \mathbb{G}_{N,\epsilon/N}$ for any ϵ with $0 < \epsilon < 1$. Then, for any $\delta > 0$,* **whp** *the graph $G = \bar{G} + R$ has some $S \subseteq V$ with $|S| = \frac{1}{2}\epsilon^{-1}\ln N$ and $e(S, \bar{S}) < \delta|S|$.*

Proof. Partition $V(\bar{G})$ into subcubes each containing $\ln N$ vertices. Each subcube S_i has sides of length $(\ln N)^{1/d}$, and, for any constant δ and N sufficiently large, $e_{\bar{G}}(S_i, \bar{S}_i) = \mathcal{O}\left((\ln N)^{(d-1)/d}\right) < \delta \ln N$. $\square$

On the other hand, if $\bar{G}$ is a graph such that all small partitions satisfying the expansion condition, then Theorem 2 is sufficient to show that G is an expander

whp. For example, if $\bar{G}$ consists of 2 expander graphs, each on $n/2$ disjoint vertices, that are joined by a single edge, then G will be an expander **whp**.

The proof of Theorem 4 goes through without modification to show that the model studied by Watts and Strogatz (SW_1 with $\bar{G}$ a k-connected cycle) is not an expander for any δ if k is any constant. On the other hand, when $k \gg \ln n$ (as in the original Watts-Strogatz specification), **whp** every vertex has at least 1 edge randomly rewired, so it follows from Theorem 1 that the resulting graph is an expander **whp**.

4   Conditions for Expansion in Kleinberg's Small-World Graph

In SW_2, when $r = 0$ and $q = 1$, this is exactly the case treated in Theorem 1. Making q larger only increases the number of edges across any cut, so any connected base graph leads to an expander when $r = 0$.

For $r > 0$, the proof of expansion can proceed as in the proof of Theorem 1, provided that there is a upper-bound on $\Pr[e_{v,i} \in S]$ with any constants $C > 0$ and $\epsilon > 0$ of the form:

$$\text{for any } v \in V \text{ and } S \subseteq V \text{ with } |S| = s, \ \Pr[e_{v,i} \in S] \leq C \left(\frac{s}{n}\right)^{\epsilon}.$$

When the metric is the ℓ_1 norm on the lattice $[n]^d$, such a bound exists for any $r < d$:

Theorem 6. *Let $V = [n]^d$, and let $d(u,v) = \sum_{i=1}^{d} |u_i - v_i|$. Then, for any r with $0 \leq r < d$, and for any $v \in V$ and $S \subseteq V$ with $|S| = s$, $\Pr[e_{v,i} \in S] \leq C \left(\frac{s}{n}\right)^{d-r-1}$.*

The proof is a direct calculation and is omitted due to lack of space.

The upper-bound on r given in this bound is tight, and when $r = d$, the resulting graph is not an expander.

Theorem 7. *For $\bar{G}$ an $n \times n$ grid, $d(\cdot,\cdot) = d_1(\cdot,\cdot)$, and $r \geq 2$, Kleinberg's small-world graph is not rapidly mixing **whp**.*

Proof. To verify the theorem, consider the set $S = \{(x,y) : x + y \leq k\}$, where $k = n/\ln n$, and calculate an upper-bound on the expected number of random edges between S and $\bar{S}$. This calculation can be simplified by considering sets $S_\ell = \{(x,y) \in V(\bar{G}) : x + y = \ell\}$. For any i and j with $i \leq k \leq j$,

$$\sum_{v \in S_i} \sum_{w \in S_j} d_1(v,w)^{-2} \leq |S_i| \left((j-i)\frac{1}{(j-i)^2} + 2 \sum_{\ell=1}^{|S_j|-(j-i)} \frac{1}{(j-i+2\ell)^2} \right)$$

$$= i \left(\frac{1}{j-i} + 2 \sum_{\ell=1}^{i} \frac{1}{(j-i+2\ell)^2} \right)$$

$$\leq 2i \left(\frac{1}{j-i} \right).$$

Also, for any $v \in V(\bar{G})$,

$$\sum_{w \neq v} d_1(v, w)^{-2} = \Theta(\ln n).$$

Thus, an upper-bound on the expected number of random edges between S and $\bar{S}$ is given by the following

$$E[e_R(S, \bar{S})] \leq \sum_{i=1}^{k} \sum_{j=k+1}^{n} \sum_{v \in S_i} \sum_{w \in S_j} q \left(\frac{d_1(v, w)^{-2}}{\sum_{u \neq v} d_1(v, u)^{-2}} + \frac{d_1(w, v)^{-2}}{\sum_{u \neq w} d_1(w, u)^{-2}} \right)$$

$$\leq (2q)\Theta\left((\ln n)^{-1}\right) \sum_{i=1}^{k} \sum_{j=k+1}^{n} 2i \left(\frac{1}{j-i} \right)$$

$$\leq (4qk)\Theta\left((\ln n)^{-1}\right) \sum_{i=1}^{k} (H_n - H_i)$$

$$= (4qk)\Theta\left((\ln n)^{-1}\right) (k + k(H_n - H_k))$$

$$= \Theta\left(\frac{k^2 \ln \ln n}{\ln n} \right).$$

Since $e_{\bar{G}}(S, \bar{S}) = \mathcal{O}(k)$, Markov's inequality shows that for any constant δ with $\delta > 0$, **whp** $e(S, \bar{S}) \leq \delta |S|$.   □

5   Conclusion

It is necessary to conclude that the expansion of a randomly perturbed graph depends on the base graph and the perturbation, and even seemingly similar perturbations can produce vastly different results. Although adding a random 1-out makes any connected graph an expander **whp**, such a simple statement is impossible for the gentler perturbation of adding a random $\mathbb{G}_{n,\epsilon/n}$. This is not a bad thing, however, as empirical observations show that among complex networks in the real world, some are expanders and others are not.

In the case of the small-world models of Watts and Strogatz and of Kleinberg, the difference in the distributions is quite subtle. Generally Kleinberg's model is viewed as a strict generalization of Watts and Strogatz's, but in the context of expansion, the models are actually just different.

It is a pleasant surprise that Kleinberg's model stops being an expander exactly at the point where it becomes possible to find short paths with a decentralized algorithm. Perhaps the expansion threshold and existence of decentralized algorithms are fundamentally related in some way. But more likely not.

Acknowledgments

Thanks to Juan Vera, Haifeng Yu, and Phil Gibbons for helpful discussions.

References

1. Flaxman, A.D., Frieze, A.M.: The diameter of randomly perturbed digraphs and some applications. In: Jansen, K., et al. (eds.) RANDOM 2004 and APPROX 2004. LNCS, vol. 3122, pp. 345–356. Springer, Heidelberg (2004)
2. Bollobás, B., Chung, F.R.K.: The diameter of a cycle plus a random matching. SIAM J. Discrete Math. 1(3), 328–333 (1988)
3. Bohman, T., Frieze, A., Martin, R.: How many random edges make a dense graph Hamiltonian? Random Structures Algorithms 22(1), 33–42 (2003)
4. Bohman, T., Frieze, A., Krivelevich, M., Martin, R.: Adding random edges to dense graphs. Random Structures Algorithms 24(2), 105–117 (2004)
5. Krivelevich, M., Sudakov, B., Tetali, P.: On smoothed analysis in dense graphs and formulas. Random Structures Algorithms 29(2), 180–193 (2006)
6. Spielman, D., Teng, S.-H.: Smoothed analysis of algorithms: Why the simplex algorithm usually takes polynomial time. In: Proceedings of the Thirty-Third Annual ACM Symposium on Theory of Computing, New York, pp. 296–305. ACM Press, New York (2001)
7. Andersen, R., Chung, F., Lu, L.: Analyzing the small world phenomenon using a hybrid model with local network flow. In: Leonardi, S. (ed.) WAW 2004. LNCS, vol. 3243, pp. 19–30. Springer, Heidelberg (2004)
8. Kleinberg, J.: The small-world phenomenon: An algorithm perspective. In: Proceedings of the Thiry-Second Annual ACM Symposium on Theory of Computing, New York, pp. 163–170. ACM Press, New York (2000)
9. Benjamini, I., Berger, N.: The diameter of long-range percolation clusters on finite cycles. Random Structures Algorithms 19(2), 102–111 (2001)
10. Coppersmith, D., Gamarnik, D., Sviridenko, M.: The diameter of a long-range percolation graph. Random Structures Algorithms 21(1), 1–13 (2002)
11. Biskup, M.: On the scaling of the chemical distance in long-range percolation models. Ann. Probab. 32(4), 2938–2977 (2004)
12. Watts, D.J., Strogatz, S.H.: Collective dynamics of "small-world" networks. Nature 292, 440–442 (1998)
13. Cheeger, J.: A lower bound for the smallest eigenvalue of the Laplacian. In: Problems in analysis (Papers dedicated to Salomon Bochner, 1969), pp. 195–199. Princeton Univ. Press, Princeton (1970)
14. Alon, N., Milman, V.D.: λ_1, isoperimetric inequalities for graphs, and superconcentrators. J. Combin. Theory Ser. B 38(1), 73–88 (1985)
15. Alon, N.: Eigenvalues and expanders. Combinatorica 6(2), 83–96 (1986) (Theory of computing (Singer Island, Fla., 1984))
16. Sinclair, A., Jerrum, M.: Approximate counting, uniform generation and rapidly mixing Markov chains. Inform. and Comput. 82(1), 93–133 (1989)
17. Lawler, G.F., Sokal, A.D.: Bounds on the L^2 spectrum for Markov chains and Markov processes: A generalization of Cheeger's inequality. Trans. Amer. Math. Soc. 309(2), 557–580 (1988)
18. Füredi, Z., Komlós, J.: The eigenvalues of random symmetric matrices. Combinatorica 1(3), 233–241 (1981)
19. Friedman, J., Kahn, J., Szemerédi, E.: On the second eigenvalue of random regular graphs. In: Proceedings of the Twenth-First Annual ACM Symposium on Theory of Computing, pp. 587–598. ACM Press, New York (1989)
20. Alon, N., Kahale, N.: A spectral technique for coloring random 3-colorable graphs. SIAM J. Comput. 26(6), 1733–1748 (1997)

21. Friedman, J.: A proof of Alon's second eigenvalue conjecture. In: Proceedings of the Thirty-Fifth Annual ACM Symposium on Theory of Computing, pp. 720–724. ACM Press, New York (2003)
22. Feige, U., Ofek, E.: Spectral techniques applied to sparse random graphs. Random Structures and Algorithms 27(2), 251–275 (2005)
23. Faloutsos, M., Faloutsos, P., Faloutsos, C.: On power-law relationships of the internet topology. In: SIGCOMM '99: Proceedings of the conference on Applications, technologies, architectures, and protocols for computer communication, pp. 251–262. ACM Press, New York (1999)
24. Monasson, R.: Diffusion, localization and dispersion relations on "small-world" lattices. The European Physical Journal B 12, 555–567 (1999)
25. Chung, F., Lu, L., Vu, V.: Eigenvalues of random power law graphs. Ann. Comb. 7(1), 21–33 (2003)
26. Mihail, M., Papadimitriou, C., Saberi, A.: On certain connectivity properties of the internet topology. In: FOCS 2003: Proceedings of the 44th Annual IEEE Symposium on Foundations of Computer Science, Washington, DC, USA, p. 28. IEEE Computer Society Press, Los Alamitos (2003)
27. Flaxman, A., Frieze, A., Fenner, T.: High degree vertices and eigenvalues in the preferential attachment graph. Internet Math. 2(1), 1–19 (2005)
28. Blandford, D.K., Blelloch, G.E., Kash, I.A.: Compact representations of separable graphs. In: Proceedings of the Fourteenth Annual ACM-SIAM Symposium on Discrete Algorithms (Baltimore, MD, 2003), pp. 679–688. ACM Press, New York (2003)
29. Estrada, E.: Spectral scaling and good expansion properties in complex networks. Europhysics Letters 73(4), 649–655 (2006)
30. Flaxman, A., Frieze, A.M., Vera, J.: A geometric preferential attachment model of networks. In: Leonardi, S. (ed.) WAW 2004. LNCS, vol. 3243, pp. 44–55. Springer, Heidelberg (2004)
31. Alon, N., Spencer, J.H.: The probabilistic method. Wiley-Interscience Series in Discrete Mathematics and Optimization, p. 301. Wiley-Interscience [John Wiley & Sons], New York (2000) With an appendix on the life and work of Paul Erdős
32. Janson, S., Łuczak, T., Rucinski, A.: Random graphs. Wiley-Interscience Series in Discrete Mathematics and Optimization. Wiley-Interscience, New York (2000)

Web Structure in 2005

Yu Hirate, Shin Kato*, and Hayato Yamana

Dept. of Computer Science, Waseda University, 3-4-1 Okubo Shinjuku-ku Tokyo,
Japan
{hirate,kato,yamana}@yama.info.waseda.ac.jp

Abstract. The estimated number of static web pages in Oct 2005 was
over 20.3 billion, which was determined by multiplying the average number of pages per web server based on the results of three previous studies,
200 pages, by the estimated number of web servers on the Internet, 101.4
million. However, based on the analysis of 8.5 billion web pages that we
crawled by Oct. 2005, we estimate the total number of web pages to be
53.7 billion. This is because the number of dynamic web pages has increased rapidly in recent years. We also analyzed the web structure using
3 billion of the 8.5 billion web pages that we have crawled. Our results
indicate that the size of the "CORE," the central component of the bow
tie structure, has increased in recent years, especially in the Chinese and
Japanese web.

1 Introduction

As of Oct. 2005, the number of static web pages was estimated at over 20.3
billion. This number was calculated by multiplying the estimated average number
of web pages per web server, 200[1][2][3], by the number of web servers in Oct.
2005, 101,435,253[4]. However, based on our crawling status by Oct. 2005[5],
we estimate the number of web pages, including both static and dynamic web
pages, in Oct. 2005 to be about 53.7 billion. This discrepancy may be due to the
increase in number of dynamic web pages generated by CGI, *etc.*

In 1999, Broder *et al.* analyzed the graph structure of the web, called the web
structure, from the set of web pages crawled in that year[6]. In this previous report, about 90% of web pages belonged to connected components, and about 28%
of web pages belonged to SCC (=strongly connected components)[6]. Inspired
by Broder's web structure, several researchers investigated the web structure
based on their crawling web data. For example, in 2003, Lie *et al.* analyzed the
structure based on the set of web pages crawled from China. They reported that
the percentage of SCC was much larger than that in Broder's web structure[8].
However, there have been no analyses of the web structure based on recent web
pages from all over the world. Here, we report the web structure computed from
3 billion web pages crawled between Jan. 2004 and Oct. 2005.

The remainder of this paper is organized as follows. In section 2, we describe
the e-Society Project[5] funded by the Japanese government. We analyzed the

---

* Currently working at Mitsubishi Electric Corporation.

W. Aiello et al. (Eds.): WAW 2006, LNCS 4936, pp. 36–46, 2008.

web structure based on the data of web pages crawled by the e-Society Project. In section 3, we report the estimated number of web pages. In section 4, we show examples of related work with respect to analysis of the web structure. In section 5, we report the results of analysis of the web structure based on our crawled web pages. Section 6 presents a summary of our work.

2   The e-Society Project[5]

The e-Society project "Technologies for the Knowledge Discovery from the Internet" is one of the projects of the Ministry of Education, Culture, Sports, Science, and Technology, Japan. The project contractor is Waseda University. The project aims (1) to gather all web pages in the world efficiently and (2) to discover some type of knowledge by applying data mining techniques. To achieve these goals, we are now gathering web pages from all over the world. As described in section 1, we used the data of pages crawled as part of this project to analyze the web structure in 2005.

2.1   Crawling Status

We began gathering web pages in Jan. 2004 with 30 CPUs in 3 different locations in Japan. We added 20 CPUs in Jan. 2005 and another 30 CPUs in Oct. 2005. Currently, our crawling system has the capability to gather up to 35 million web pages per day. By Oct. 2006, we had gathered over 14 billion web pages from all over the world.

3   How Many Web Pages Are There?

As the most basic analysis, we estimated the number of web pages on the web. Conventional research indicates that each web host has an average of about 200 web pages[1][2][3]. Netcraft, a company from the UK, investigates the number of web servers on the whole web and publishes their results every month on their home page. According to the Netcraft report of Nov. 2005[4], the number of web servers was estimated as 101,435,253. By multiplying these two numbers -$i.e.$, $200 \times 101,435,253$- we can estimate the total number of web pages as about 20.3 billion.

However, by Oct. 2005, we had gathered 8,507,237,370 web pages from 16,035,801 web servers, indicating that the average number of web pages per host is about $8,507,237,370/16,035,801 \simeq 530$. Multiplying this figure by the number of web servers reported by Netcraft yields an estimate of the total number of web pages all over the world of about 53.7 billion. This discrepancy may be due to the recent rapid increase in number of dynamic web pages, such as CGI pages based on databases, Blogs, Portal Sites, and EC sites.

Note that this analysis differs from estimating the number of web pages indexed by search engines. Bharat et $al.$[9], Henzinger et $al.$[10] Vaughan et $al.$[11] and Bar-Yossef et $al.$[12] investigated the relative size of several search engines' indexes in 1997, 1998, 2004 and 2006, respectively. These works focus on the relative size of indexed web pages by different search engines. On the other hand, our analysis estimates the size of actual web, regardless of whether each web page is indexed or not.

4   Related Works

In this section, we present related work with respect to web structure[6].

4.1   Applying Graph Theory to Web Link Structure

The web has a hyperlinked structure, which was first introduced by Broder *et al.* [6]. When we consider pages as vertexes and hyperlinks as edges, then it is possible to regard the web link structure as a directed graph. Broder *et al.* focused on this property, and analyzed the web link structure from the viewpoint of graph theory[6]. Inspired by their approach, several researchers analyzed the web structure based on their own web data. The remainder of this section describes these conventional studies.

4.2   Graph Structure in the Web[6]

Broder *et al.* analyzed the whole web structure in 1999 based on 200 million web pages with 1,500 million hyperlinks[6]. They reported that about 90% of web pages belong to connected components and the structure is similar in shape to a bow tie, as shown in Fig. 1.

Broder *et al.* defined four types of component in the web structure, IN, CORE, OUT, and TENDRILS[6]:

- CORE is defined as SCC(=Strongly Connected Component).
- IN is defined as the set of web pages that have paths to SCC but do not have paths from SCC.
- OUT is defined as the set of web pages that have paths from SCC but do not have paths to SCC.
- TENDRIL is defined as the set of web pages that do not have paths either to or from SCC.

As shown in Table 1, the web structure in 1999 consisted of 28% SCC and 21% IN, OUT, and TENDRILS.

4.3   Structural Properties of the African Web[7]

Boldi *et al.* analyzed the African web structure in 2002. The dataset consisted of 2 million web pages gathered from 2,500 African hosts. They reported that the African web differed in shape from the bow tie structure of the web as a whole. As shown in Fig. 2, the African web structure contained multiple OUT components, but had no IN components. However, the dataset of the African web structure was small compared to those of other web structures, and this may have been responsible for the difference in shape.

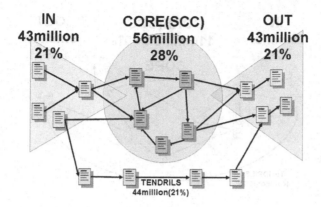

Fig. 1. Bow-tie Structure of the Web in 1999[6]

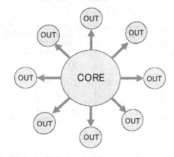

Fig. 2. African Web Structure in 2002[7]

4.4   China Web Graph Measurements and Evolution[8]

Lie *et al.* analyzed the Chinese web structure in 2003 based on 140 million web pages with 4,300 million hyperlinks[8]. They reported that the Chinese web structure was bow tie-shaped, as shown in Fig. 3. However, as shown in Table 1, the percentage of CORE in the Chinese web structure in 2003 was much larger than that of in Broder's whole web structure in 1999. The authors concluded that the large percentage of CORE in the Chinese web structure was a phenomenon specific to the Chinese web.

5   Web Structure in 2005

As we described in section 4, conventional studies have revealed different properties of the web structure. Broder's web structure was based on the web data in 1999. Boldi's African web structure and Lie's Chinese web structure were based on parts of the whole web. To determine the current shape of the web structure, an updated web structure is needed. We analyzed the structure based on data

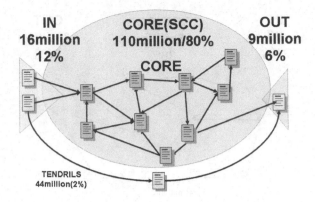

Fig. 3. Chinese Web Structure in 2003[8]

Table 1. The percentages of components of bow-tie structures[6][8]

WebGraph	CORE(SCC)	IN	OUT	TENDRILS	DISCONNECTED
Web in 1999[6]	56 million (28%)	43 million (21%)	43 million (21%)	44 million (21%)	17 million (8%)
Chinese Web in 2003[8]	112 million (80%)	16.5 million (12%)	9 million (6%)	1 million (0.7%)	1 million (0.7%)

consisting of web pages gathered as part of the e-Society project. In this section, we report the results for the current web structure.

We analyzed the whole web structure, web structures by TLD (=Top Level Domain), and web structures by language based on 3,207,736,427 web pages[1] gathered between Jan. 2004 and Jul. 2005. These web pages were gathered from all over the world and their languages were detected automatically using the Basis Technology Rosette Language Identifier[13].

The reminder of this section is as follows. In section 5.1, we introduce our analytical strategy for computation, *i.e.*, host level reduction. In section 5.2, we describe the properties of our dataset. Then, we report the results of the whole web structure, web structures by TLD, and web structures by language in sections 5.3, 5.4, and 5.5, respectively.

5.1 Host Level Reduction

Our analysis was based on host level analysis as described below:

– Pages in the same host are considered as one vertex.
– Hyperlinks to other hosts are considered as edges.

---

[1] Our crawler gathers web pages from the top page of each host, and follows links up to 15 stratums.

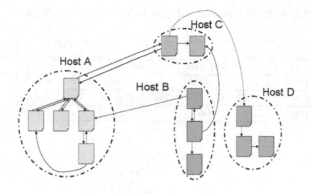

Fig. 4. Raw Link Data

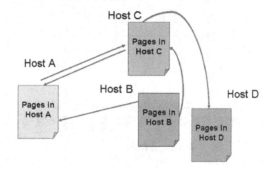

Fig. 5. Preprocessed Link Data

To analyze the host level web structure, we applied the following preprocessing steps to a raw link dataset before computing the web structure.

1. Extract hosts from a raw link dataset to generate a host list.
2. Group web pages in the same host.
3. Extract links that connect two different hosts, called inter-host links, from the raw link dataset.

We call these preprocessing steps "host level reduction." For example, when we have a raw link dataset as shown in Fig. 4, preprocessed data will be as shown in Fig. 5.

5.2   Dataset Properties

We used the data of web pages crawled as part of the e-Society project. The TLD distribution and language distribution of the dataset are shown in Fig. 6 and Fig. 7, respectively. Similar to the actual web, our data collection consisted of a large proportion of ".com" domains (Fig. 6) and web pages in English (Fig. 7).

Table 2. Components of Whole Web Structure in 2005

	CORE	IN	OUT	Other
Number of hosts	624,173	147,794	621,788	119,770
Percentage of hosts	41%	10%	S 41%	8%
Number of pages	2,102,971,321	633,530,035	346,251,616	124,983,455
Percentage of pages	65%	20%	11%	4%

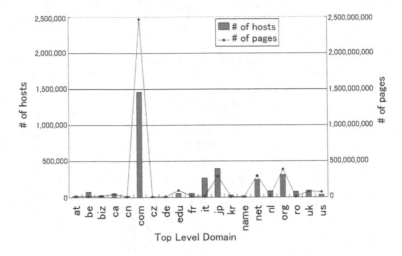

Fig. 6. TLD Distribution of the Dataset

However, as we gathered web pages from ".jp" and ".com" domain lists, our dataset was biased toward ".jp" and ".com" domains.

Our raw link dataset consisted of 3,208,139,905 web pages (vertexes) and 93,397,065,743 links (edges). After applying host level reduction preprocessing, the preprocessed link data consisted of 1,719,134 hosts (vertexes) and 91,084,879 inter-host links (edges).

In this analysis, we have not discarded duplicated web pages in a host nor among hosts. However, because of the host level reduction, which is described in section 5.1, web pages in the same host are considered as a vertex. Due to this, our result turns to be the same even if duplicated web pages in the same host are aggregated. On the other hand, when duplicated web pages exist among hosts, they are considered as multiple vertexes.

5.3   The Whole Web Structure in 2005

Fig. 8 and Table 2 show the results of the whole web structure in 2005. As shown in Fig. 8 and Table 2, the percentage of the CORE component had become larger than that in Broder's whole web structure in 1999. Although Lie *et al.* concluded

Table 3. Components of web structures by TLD in 2005

TLD	CORE	IN	OUT	Other
.com	53.65%	19.73%	22.25%	4.37%
.jp	26.46%	1.77%	71.32%	0.46%
.de	0.25%	0.05%	78.36%	21.34%
.edu	0.05%	0.00%	14.44%	85.51%
.fr	0.01%	0.02%	25.33%	74.63%
.it	0.11%	0.04%	0.04%	99.81%
.kr	0.00%	0.00%	1.09%	98.91%
.net	0.52%	0.17%	35.42%	63.89%
.org	0.61%	0.38%	64.25%	34.76%
.ru	0.77%	0.05%	0.49%	98.70%

that the increase in the CORE component percentage was a pattern specific to the Chinese web, this phenomenon was found not only in the Chinese web, but also in the whole web in the present study.

5.4 Web Structures by TLD

Table 3 shows the results of web structures by TLD. As shown in Table 3, there were no large CORE components in web structures of all TLD, excluding the .com and .jp domains. Even in the .jp domain web structure, the percentage of CORE component was smaller than that in Broder's whole web structure in 1999. These results indicate that the web cannot be divided with regard to TLD.

5.5 Web Structures by Language

Table 4 shows the results of web structure by language. As shown in Table 4, Chinese and Japanese language web structures have large percentages of CORE

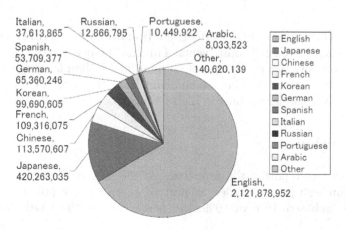

Fig. 7. Language Distribution of the Dataset

Table 4. Components of web structures by Language in 2005

TLD	CORE	IN	OUT	Other
Chinese	76.88%	9.98%	10.57%	2.57%
Japanese	71.05%	25.85%	2.54%	0.56%
English	66.90%	9.04%	16.44%	7.62%
Spanish	64.93%	5.30%	23.60%	6.16%
French	61.85%	9.23%	20.65%	8.27%
Arabic	61.43%	10.20%	18.59%	9.78%
Korea	54.32%	17.07%	19.36%	9.25%
Russian	35.76%	18.20%	18.35%	27.69%
Portuguese	26.60%	4.94%	42.18%	26.28%
German	26.61%	8.16%	42.18%	23.05%
Italian	23.67%	17.10%	29.54%	29.69%
Other	7.24%	1.98%	9.32%	81.47%

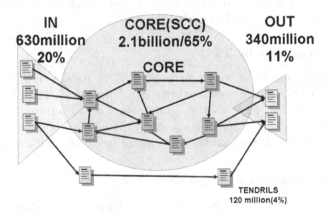

Fig. 8. Web structure in 2005

components. On the other hand, German and Italian language web structures have small percentages of CORE components. This may be because our web data collection had low coverage in the case of German and Italian language web pages. Note that in the case of the Chinese language web structure, the percentages of components were similar to those in Lie's Chinese web graph in 2003.

6   Conclusion

In this paper, we reported the properties of the web structure in 2005 based on 3.2 billion web pages crawled as part of the e-Society project. Compared to Broder's web structure in 1999, the percentage of the CORE component increased from 28% to 65%. Lie *et al.* concluded that the large percentage of CORE component is a phenomenon specific to the Chinese web structure. However, our analysis showed that the increase in CORE component has occurred in the whole web.

We also analyzed web structures by TLD and by language. By comparing the two types of web structure, we confirmed that the web cannot be divided by TLD, but can be divided by written language.

6.1 Future Work

Since our analysis is based on the crawled web pages which have been crawled by our original crawler following hyper-links from seed URLs, our dataset might loss some web pages which should be categorized as IN component. Then this might result in the decreased percentage of IN component to some extent. To solve this, one can use indexes of search engines, and internet archive, such as WebBase[14][15] to add the fraction of IN component to the dataset. As a future work, we update the web structure with solving such a IN component problem.

Acknowledgements

This research was funded in part by both "e-Society: the Comprehensive Development Foundation Software" of MEXT (Ministry of Education, Culture, Sports, Science, and Technology) and "21st Century COE Programs: ICT Productive Academia" of MEXT.

References

1. Lawrence, S., Giles, C.L.: Searching the World Wide Web. Science 280(5360), 98–100 (1998)
2. Lawrence, S., Giles, C.L.: Accessibility of Information on the Web. Nature, 400, 107–109 (1999)
3. Institute for Information and Communications Policy, Statistics Investigation Report for contents on the World-Wide Web (2004),
   http://www.soumu.go.jp/iicp/chousakenkyu/seika/houkoku.html
4. Netcraft: Web Server Survey (November 2006),
   http://news.netcraft.com/archives/2006/11/01/
   november_2006_web_server_survey.html
5. e-Society Project,
   http://www.yama.info.waseda.ac.jp/~yamana/e-society/index_eng.htm
6. Broder, A., Kumar, R., Maghoul, F., Raghavan, P., State, R., Tomkins, A., Wiener, J.: Graph structure in the web. In: Proc. of 9th World Wide Web Conf., pp. 309–320 (2000)
7. Boldi, P., Codenotti, B., Santini, M., Vigna, S.: Structural Properties of the African Web. In: Poster Proc. of 11th World Wide Web Conf., (2002)
8. Lie, G., Yu, Y., Han, J., Xue, G.: China web graph measurements and evolution. In: Zhang, Y., Tanaka, K., Yu, J.X., Wang, S., Li, M. (eds.) APWeb 2005. LNCS, vol. 3399, Springer, Heidelberg (2005)

9. Bharat, K., Broder, A.: A Technique for Measuring the Relative Size and Overlap of Public Web Search Engines. Journal of Computer Networks and ISDN Systems 30(1-7), 379–388 (1998)
10. Henzinger, M., Heydon, A., Mitzenmacher, M., Najork, M.: Measuring Index Quality using Random Walks on the Web. In: Proc. of 8th World Wide Web Conf., pp. 213–225 (1999)
11. Vaughan, L., Thelwall, M.: Search Engine Coverage Bias: Evidence and Possible Causes. Journal of Information Processing and Management 40(4), 693–707 (2004)
12. Bar-Yossef, Z., Gurevich, M.: "Random Sampling from a Search Engine's Index. In: Proc. of 15th World Wide Web Conf., pp. 367–376 (2006)
13. Basis Technology Rosette Language Identifier,
    http://www.basistech.com/language-identification/
14. Hirai, H., Raghavan, S., G-Molina, H., Paepcke, A.: Webbase: A repository of the Web. In: Proc. of 9th World Wide Conf., pp. 277–293 (2000)
15. The Stanford WebBase Project,
    http://dbpubs.stanford.edu:8091/~testbed/doc2/WebBase/

Local/Global Phenomena in Geometrically Generated Graphs

Ross M. Richardson[1,2]

[1] Department of Mathematics
University of California, San Diego
La Jolla, CA 92093-0112, USA
[2] Center for Combinatorics. LPMC
Nankai University, Tianjin 300017
People's Republic of China
rmrichardson@math.ucsd.edu

Abstract. We study a geometric random tree model $\mathcal{T}_{\alpha,n}$ which is a variant of the FKP model proposed in [1]. We choose vertices $v_1, \ldots, v_n$ in some convex body uniformly and fix a point $\mathfrak{o}$. We then build our tree inductively, where at time t we add an edge from v_t to the vertex in $v_1, \ldots, v_{t-1}$ which minimizes $\alpha \|v_t - v_i\| + \|v_i - \mathfrak{o}\|$ for $i < t$, where $\alpha > 0$. We categorize an edge $v_i \rightarrow v_j$ in this graph as local or global depending on the edge length relative to the distance from v_i to $\mathfrak{o}$. It is shown that for α bounded away from 1 either all edges are local or all are global a.a.s. However, as $\alpha \rightarrow 1$ we show that in fact the number of local and global edges are asymptotically balanced.

1   Introduction

Consider the problem of providing telephone service to some central hub. Each customer has some given position in the plane, and thus a distance from the hub. If we are allowed to extend a single connection from each new customer to an existing customer, then choosing the nearest neighbor clearly optimizes the amount of new wire we have to string. On the other hand, connecting to an existing customer who is farther away from the hub than our new customer may lead to attenuated service, and hence we may wish to choose a customer who is closer to the hub (in the Euclidean sense, say). If we weight these two costs and choose a customer which optimizes our total cost function, the behavior will clearly depend on the relative weighting given to each cost.

We shall construct a simple geometric tree model motivated by the above example. We shall say that a customer is linked by a *local edge* (resp. *global edge*) if the edge length is short (resp. long) *relative to the distance between the customer and the hub*. In this way we obtain a meaningful description of the local behavior which respects the length scale of each vertex. The above example shows that, depending on how we choose to weight edge costs, both completely local and completely global behavior are possible. In this paper, we shall quantify these ideas, and in particular we shall look at the transition from global to local behavior based on relative costs.

W. Aiello et al. (Eds.): WAW 2006, LNCS 4936, pp. 47–58, 2008.
© Springer-Verlag Berlin Heidelberg 2008

2   Related Work

Network models which encapsulate both local and global structure have been investigated for some time. One of the earliest such papers is [2], in which the authors analyze the union of an n-cycle and a random matching, showing that it has both a linear number of edges and small diameter.

From an algorithmic perspective, Kleinberg in [3] proposed a simple local/global network model consisting of an $n \times n$ planar grid, to which one adds a single random edge at each vertex. The edge is chosen with probability proportional to some power of the inverse distance. He demonstrates that only for power 2 can an algorithm find short paths given only local information; for other exponents the random component is either too local or too random. In a similar vein, the authors of [4] show that for an arbitrarily populated grid model in which link probabilities are determined via local density (sparse regions have higher link probability than dense regions), computable short paths exist. The authors of [5] replace the local grid with a graph that satisfies prescribed local flow constraints, and the global graph with a power-law $G(\mathbf{w})$ random graph (see [6]). In this way they obtain both local clustering and small diameter, and they further provide an algorithm for separating the local and global components.

The model we present here is a variant of that originally proposed in [1] (the FKP model). In their model, a random tree is formed in the unit square, where each new vertex v_t is attached to the prior vertex v_i that minimizes $\alpha \left\| v_i - v_t \right\| + d(v_t, 0)$, where here $\left\| \cdot \right\|$ denotes the Euclidean distance and $d(\cdot, 0)$ the graph theoretic distance to the root. This model is an example of the *Heuristically Optimized Tradeoff* paradigm, popularized in [7], characterized by optimization of a random hazard. They show that for $\alpha = o(\sqrt{n})$ the graph is essentially star-like while for $\alpha = \omega(\sqrt{n})$ the graph behaves as a random recursive tree. While this model was originally investigated with respect to power-law behavior, it was shown in [8] that in these regimes a power law can only exists in a vanishingly small fraction of the tail, but both [1] and [8] leave unanalyzed the case of $\alpha = \Theta(\sqrt{n})$.

Finally, geometric random graph models as such have been studied for over a decade, the most studied being the disc model in which two points of some random point process are linked if they are of distance less than r (see [9]). Random linkages in point processes have existed in the infinitary setting for much longer in the percolation literature [10]. The use of more delicate models in complex network modeling is more recent. The authors of [11] consider n random points on the sphere, where each vertex connects to a fixed number k vertices in a neighborhood of radius r about the vertex. Their model generates a power-law degree distribution and can be shown to have small separators.

3   Definitions and Model

We define a random graph model $\mathcal{T}_\alpha$ ($\mathcal{T}_{\alpha,n}$ when we wish to stress the dependence on n) with positive parameter α. We denote by K some compact, convex set in

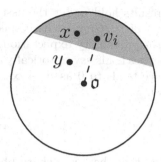

Fig. 1. The shaded region represents $H(v_i, 1/4)$ where K is a sphere. Note that x and y are 1/4 local and global, respectively.

$\mathbb{R}^d$ and o a fixed point in K. Without loss of generality, we assume the volume of K to be one.

For a given natural number n, we construct $\mathcal{T}_{\alpha,n}$ as follows: The vertices of $\mathcal{T}_{\alpha,n}$, denoted by $V_n = \{v_1, \ldots, v_n\}$, are chosen independently and uniformly in K.[1] For each vertex $v_i, i = 1, \ldots, n$, associate to it a function

$$\phi_i(x) = \alpha \|v_i - x\| + \|x - \mathsf{o}\|, \tag{1}$$

where here $\|\cdot\|$ is the Euclidean length. For the model $\mathcal{T}_{\alpha,n}$, we associate to each vertex $v_i, i = 1, \ldots, n$, a unique edge e_i with source v_i and target v_{t_i} given by

$$v_{t_i} := \operatorname{argmin}_{j < i} \phi_i(v_j).$$

We shall assume in the sequel that such a minimizer is unique, as this happens with probability one. One may also define an offline version in which all vertices are revealed at once and the selection rule $\operatorname{argmin}_{j \neq i} \phi_i(v_j)$ is used to determine the target of edge e_i. Unlike $\mathcal{T}_\alpha$, however, this later model is not in general a tree.

Finally, for an edge e_i with target v_j, we call e_i β−local (or $(1 - \beta)$-global) for $0 \leq \beta \leq 1$ if the Euclidean edge length of the orthogonal projection of $\overline{v_i v_j}$ onto the segment $\overline{\mathsf{o} v_i}$ is at most $\beta \|v_i - \mathsf{o}\|$. Equivalently, the target of the edge is contained in the halfspace

$$H(v_i, \beta) := \left\{ x \mid (x - \mathsf{o}) \cdot (v_i - \mathsf{o}) \geq (1 - \beta) \|v_i - \mathsf{o}\|^2 \right\}.$$

If not otherwise stated we take $\beta = \frac{1}{2}$. See Fig. 1. We shall concern ourselves with the fraction $\rho(\beta)$ of β-local edges in $\mathcal{T}$.

---

[1] As is usual in the theory of random graphs, we shall adopt the point of view that $\mathcal{T}_{\alpha,n}$ and $\mathcal{T}_{\alpha,m}$ are constructed on the same probability space such that $V_n \cap V_m = V_n$ if $n \leq m$. As the vertices completely determine the graph, the subgraph of $\mathcal{T}_{\alpha,m}$ induced by V_n is thus $\mathcal{T}_{\alpha,n}$, and hence we view $\mathcal{T}$ as being built one vertex and edge at a time.

Finally, all asymptotic results hold under the assumption $n \to \infty$, and we use the Landau notation $O(\cdot), o(\cdot), \Omega(\cdot)$ etc. with respect to this assumption. We denote by $\mathbf{P}, \mathbf{E}, \mathbf{Var}, \mathbf{Cov}$, the probability, expectation, variance, and covariance, respectively. We say a properties holds asymptotically almost surely (a.a.s) if the probability of non-occurrence tends to 0 as $n \to \infty$. The ϵ-ball about a point $p \in \mathbb{R}^d$ is denoted by $B(p, \epsilon)$.

4   Results

Our first result demonstrates that there is a sharp phase transition around the value $\alpha = 1$.

Theorem 1 (Degree Distribution). *Let K be a compact, convex set in $\mathbb{R}^d$, and o some fixed point in the interior of K. We then have the following for $T_{\alpha,n}$:*

1. *If $\alpha > \delta > 1$ for δ fixed, then there exists a constant c such that for any vertex $v_i, i = 1, \ldots, n$, we have*

$$\mathbf{P}[\deg(v_i) \geq D] = O(n \exp(-cD)). \tag{2}$$

   In particular, the maximum degree is $O(\ln n)$ a.a.s.
2. *If $\alpha < \delta < 1$ for δ fixed, then a.a.s. the number of degree one vertices is $n(1 + o(1))$, and some vertex has degree at least $\omega(n/\ln n)$.*

Remark 1. Under the assumption that $\alpha \to \infty$ sufficiently fast ($\omega(\ln n)$ say), one can show a matching exponential lower bound in (2), which follows by a modification of an argument found in [8].

The FKP model in [1] was also shown to have both star-like and exponential-tailed degree distributions based on the parameter α. However, the authors of [1] also asserted a third range of behavior ranging from the regime where α was constant to $O(\sqrt{n})$ in which the degree distribution had a power-law tail. Though consistent with the work of [1], it was subsequently shown in [8] that up to $O(\sqrt{n}/(\ln n)^4)$ this power law held only over a vanishingly small portion of the tail, and indeed the actual behavior was similar to that of part 2 of Thm. 1.

Despite the change in degree distribution, the diameter is well behaved for all parameter ranges. We have

Theorem 2 (Diameter). *For any $\alpha > 0$, the diameter of T is $\Theta(\ln n)$ a.a.s..*

Our main goal is to understand the relative number of local edges in our graph. For the case of α bounded away from 1 we can quantify edge lengths in absolute terms.

Theorem 3 (Edge Length). *We assume that $K \in \mathbb{R}^d$.*

1. *If $\alpha > \delta > 1$ for δ fixed, then there exists some constant c such that the length of e_i is at most $c(\frac{\ln n}{n})^{1/d}$ for every $i = 1, \ldots, n$, with probability tending to one. Moreover, the number of edges which exceed $(\frac{\omega(n)}{n})^{1/d}$ is $o(n)$ for any $\omega \to \infty$.*

2. *If $\alpha < \delta < 1$ for δ fixed, then all edges have one endpoint within $(\frac{\omega(n)}{n})^{1/d}$ of o a.a.s., where $\omega(n) \to \infty$ as slowly as desired.*

In particular, this gives us information on the ratio of β-local edges.

Corollary 1. *For K and o as in Thm. 3, if $\alpha > \delta > 1$ for δ fixed, then $\rho(\beta) \to 1$ almost surely. If $\alpha < \delta < 1$ for δ fixed, then $\rho(\beta) \to 0$ almost surely.*

Remark 2. In particular, we find that when α is bounded away from one the distribution of edge lengths is governed primarily by the geometry of the extreme cases, $\alpha = 0$ and $\alpha \to \infty$. Hence, as $\rho(\beta)$ is a rough measure of the local tendency of the graph, we see that the graph is entirely local or global in this case.

On the other hand, we shall see that as $\alpha \to 1$, $\rho(\beta)$ changes. If we allow β to vary with n, then we can track this relationship. A rough calculation shows:

Theorem 4. *Let $\alpha > 1$. Then if we set*

$$\beta = \omega\left((\alpha - 1)^{-\frac{d-1}{d}} n^{-1/d}\right),$$

we find $1 - \rho(\beta) = o(1)$ a.a.s..

The above is far from best possible, as we shall see from the next result, but it already yields information on the range of α for which one can find completely local behavior. A precise description of the tradeoff between α and β will appear in a forthcoming work.

Our main theorem asserts that around $\alpha = 1$, our model $\mathcal{T}_\alpha$ consists of both local and global edges of the same magnitude, as measured by our parameter $\rho(\beta)$. We work in the unit volume ball in $\mathbb{R}^2$, for simplicity.

Theorem 5 $(\alpha = 1)$. *Set $K = B(0, \pi^{-1/2}) \subseteq \mathbb{R}^2$ with $o = 0$. If $\alpha = 1$, we have*

$$\rho\left(\frac{1}{2}\right) = \frac{1}{2} + o(1),$$

with probability tending to one. Additionally, if β is fixed and $\alpha = 1 + o(n^{-2})$ then there exists constants $0 < c < c' < 1$ depending on β such

$$c < \mathbf{E}[\rho(\beta)] < c'.$$

Remark 3. The proof of Thm. 5 gives the following heuristic explanation for this behavior: for a point v_i in K with $\|v_i - o\| = l$, the edge e_i has length uniformly chosen in $[0, l]$ and is contained in the thin tube about $\overline{ov_i}$. It is thus tempting to conjecture that $\rho(x) \approx x$, but the proof of Thm. 5 does not seem to generalize to this case.

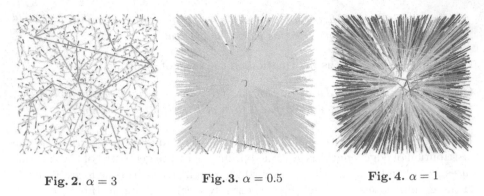

Fig. 2. $\alpha = 3$ **Fig. 3.** $\alpha = 0.5$ **Fig. 4.** $\alpha = 1$

Fig. 5. $\mathcal{T}_{\alpha,10,000}$ instances

5   Further Directions

- Though [8] shows that no reasonable power law for the FKP model exists for most ranges of α, both [8] and [1] leave substantially unexamined the cases around the phase transition, $\sqrt{n}/(\ln n)^4 \ll \alpha \ll \sqrt{n}\ln n$. However, this is the range in which the α-scaled shortest distance and the graph theoretic distance are of the same order. In the model $\mathcal{T}_\alpha$, the summands in ϕ_i are of the same order at the phase transition ($\alpha \approx 1$) as well. While the unknown constants in the FKP model make it impossible to compute examples in this range, it is easy to simulate $\mathcal{T}_1$. In such simulations we find, in examples scaling up to 10,000 vertices, that a power law holds for the *entire degree sequence*, in both the $\mathcal{T}$ and related offline model. We thus conjecture that both the $\mathcal{T}$ and FKP models actually have power-law degree distributions at the phase transitions, and Thm. 5 suggests the range of stability of this behavior for $\mathcal{T}$ (namely $\alpha = 1 + o(n^{-2})$) for $d = 2$. Further, the power law exponents appear to be decreasing functions of the dimension.
- In both $\mathcal{T}$ and the FKP model, the weight α is a fixed function of n. A. Flaxman (personal communication) suggests looking at a non-homogeneous variant of these models. To wit, for the $\mathcal{T}$ model he suggests setting

$$\phi_i(x) = \alpha(i)\,\|v_i - x\| + \|x - \mathsf{o}\|,$$

where here $\alpha(i)$ varies with i.
- Much more is left to be said about $\rho(\beta)$ for $\alpha \approx 1$. The proof of Thm. 5 presents sufficient dependence between edges such that standard sharp concentration results do not apply. Obtaining a sharper convergence result in Thm. 5, for example, will most likely require new tools.
- The results of Thm. 1 and Thm. 5 are similar to an infinite Pólya urn model studied in [12]. In this model, at each time step a ball is added to an existing urn with probability $1 - p$, else a new urn is created. If a ball is to be placed into an existing urn, then each urn is chosen with probability proportional to m^γ, where m is the number of balls in the urn. Under the regimes $\gamma > 1$,

$\gamma < 1$, and $\gamma = 1$ the bin distributions are exponential, dominated by a single bin, and power law a.a.s. Given that ϕ_i uses no explicit information about vertex degrees, it it currently unclear to what extent the degree of a vertex v_i correlates with $\|v_i - o\|$.

6 Proofs

For the sequel we focus on a proof of our main theorem, Thm. 5, in the offline version of our model, as well as the pertinent geometry underlying all of the results. An expanded treatment of the remaining results and the related offline model will appear in a forthcoming paper.

6.1 Regions of Influence

All of the subsequent analysis in this paper relies on the notion of *influence regions*. Set $\gamma > 0$. Then the set of points

$$U(p,\gamma) = \{q \mid \|o - q\| + \alpha \|q - p\| \le (\min(1,\alpha) + \gamma) \|p - o\|\}$$

forms the γ-*influence region* (or γ−region) about p. We then have:

Lemma 1 (Convexity). *The region $U(p,\gamma)$ is convex for any choice of p and $\gamma > 0$.*

Proof. Let $\mathbf{x}_1, \mathbf{x}_2 \in K$ be such that

$$\|\mathbf{x}_i - o\| + \alpha \|\mathbf{x}_i - p\| = (\min(1,\alpha) + \gamma) \|p - o\|, \qquad i = 1, 2, \qquad (3)$$

which is to say that they lie on the boundary of $U(p,\gamma)$. Let $\mathbf{z} = \lambda\mathbf{x}_1 + (1-\lambda)\mathbf{x}_2$. Then

$$\|\mathbf{z} - p\| = \|\lambda\mathbf{x}_1 + (1-\lambda)\mathbf{x}_2 - p\| \le \lambda \|\mathbf{x}_1 - p\| + (1-\lambda) \|\mathbf{x}_2 - p\|,$$

and similarly for $\|\mathbf{z} - o\|$. Thus,

$$\|\mathbf{z} - o\| + \alpha \|\mathbf{z} - p\| \le (\lambda \|\mathbf{x}_1 - o\| + (1-\lambda) \|\mathbf{x}_2 - o\|)$$
$$+ \alpha(\lambda \|\mathbf{x}_1 - p\| + (1-\lambda) \|\mathbf{x}_2 - p\|)$$
$$\text{(by (3))} = (\min(1,\alpha) + \gamma) \|p - o\|.$$

$\square$

For the special case $\alpha = 1$, the γ-region is simply an ellipse with foci o and p and with major axis length $r(1 + \gamma)/2$.

For $\alpha < 1$, as $\gamma \to \infty$ the γ region approaches that of an ellipse with foci o and p. However, for γ sufficiently small, the γ-region localizes about the point o. Specifically, the γ-region forms a convex region about o, the boundary of which is at maximum and minimum distance from o along the line through o and p. The case $\alpha > 1$ is similar, but in this case the γ-region concentrates about p. See Fig. 6.

We can further elucidate the structure of this region by computing the radii of the smallest enclosing circle and the largest inscribing circle of $U(p,\gamma)$. We summarize this as follows:

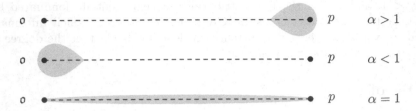

Fig. 6. γ−regions for γ small

Lemma 2. *Fix d. Let p be a point of distance r to o, and assume $\alpha \neq 1$. Then for γ sufficiently small*

$$\frac{\mathrm{Vol}(B(0,1))}{(1+\alpha)^d} \leq \frac{\mathrm{Area}(U(p,\gamma))}{(r\gamma)^d} \leq \frac{\mathrm{Vol}(B(0,1))}{|\alpha-1|^d}, \tag{4}$$

where $\mathrm{Vol}(B(0,1)) = \frac{\pi^{d/2}}{\Gamma(d/2)+1}$ denotes the volume of the unit ball in $\mathbb{R}^d$.

Proof. For $\alpha > 1$, consider the ball centered at p of minimum radius $xr, x \leq 1$ that includes $U(p,\gamma)$. As $U(p,\gamma)$ and this ball intersect along the line po, we obtain the equation $(1-x)r + \alpha xr = (1+\gamma)r$, hence $xr = \gamma r/(1-\alpha)$. The other cases are similar.   □

Thus, when $\alpha \neq 1$, the γ-region for a point p concentrates strongly about p when $\alpha > 1$ (resp. o when $\alpha < 1$). If some other point is in $U(p,\gamma)$, then we are assured that p links to a point in $U(p,\gamma)$. Hence if α is bounded away from 1, a comparison argument using Lem. 2 shows that the length of the edge from p will be of the same order as the nearest neighbor distance (resp. distance to o). These ideas can be quantified to prove Thms. 1, 3. However, if $\alpha \to 1$ observe that right hand side of (4) becomes large, reflecting the fact that the γ region becomes increasingly ellipsoidal. In this case, better estimates are necessary to prove results such as Thm. 4.

6.2   Proof of Theorem 5

We focus here on the offline model, where we make no restriction that a vertex v_i must link to some prior vertex $v_j, j < i$. We briefly discuss the modification for $\mathcal{T}_\alpha$ at the end of the proof. Our argument will be via a second moment calculation, according to the following statement (see [13]):

Lemma 3. *Let $X = X_1 + \ldots + X_n$ where X_i is an indicator for the event A_i. If $\mathbf{Var}[X] = o(\mathbf{E}[X]^2)$ then we have*

$$X = \mathbf{E}[X](1 + o(1)), \qquad a.a.s.$$

Let X_i be the indicator that e_i is a local edge, and $X = \sum_{i=1}^n X_i$ then counts the number of local edges. Denote by $E(p,t)$ the closed ellipse with foci at p and o and area t.

We shall construct parameters t_0, h_0, r_0, θ_0 depending on n, but will delay setting these for the moment. The parameter values are related by the following geometric estimate, which we assume.

Lemma 4. *Let v_i and v_j be vertices located at a radius greater than r from $\mathbf{o}$. Then the furthest point of of $E(v_i, t) \cap E(v_j, t)$ from $\mathbf{o}$ occurs at a radius at most $c_1 \frac{t^2}{r^5(1-\cos\frac{\theta}{2})}$ from $\mathbf{o}$ for some constant $c_1 > 0$, under the assumption that t, r, θ tend to zero.*

Construct the following events:

- $\mathcal{A}(v_i)$: $E(v_i, t_0) \subseteq K$,
- $\mathcal{B}(v_i)$: $|\{v_1, \ldots, v_n\} \cap (E(v_i, t_0) - B(\mathbf{o}, h_0) - B(v_i, h_0))| > 1$ and $\|v_i - \mathbf{o}\| > r_0$.
- $\mathcal{C}(v_i, v_j)$: $\angle v_i \mathbf{o} v_j > \theta_0$.

For convenience, we write $\mathcal{D}(v_i, v_j) = \mathcal{A}(v_i) \cap \mathcal{A}(v_j) \cap \mathcal{B}(v_i) \cap \mathcal{B}(v_j) \cap \mathcal{C}(v_i, v_j)$. The key to our proof is the following lemma. Note that if $\mathcal{E}$ denotes an event then $\overline{\mathcal{E}}$ denotes its compliment.

Lemma 5. *There are choices of t_0, h_0, r_0, θ_0, depending upon n, such that $h_0 \leq c\frac{t_0^2}{r_0^5(1-\cos\frac{\theta_0}{2})}$ as in Lem. 4, $\sum_{i=1}^n \mathbf{P}(\mathcal{A}(v_i) \cap \mathcal{B}(v_i)) = o(n)$ and $\sum_{i \neq j} \mathbf{P}(\overline{\mathcal{D}(v_i, v_j)}) = o(n^2)$, where the latter sum ranges over all pairs $(i, j), i \neq j$.*

Proof (Thm. 5, $\beta = 1/2$.).

Step 1: Fix i. We construct a map T_1 of the underlying probability space K^n. On the set $\overline{\mathcal{A}(v_i) \cap \mathcal{B}(v_i)}$ T_1 acts as the identity. On $\mathcal{A}(v_i) \cap \mathcal{B}(v_i)$ T_1 reflects every point in $E(v_i, t_0)$ except v_i about the axis of $E(v_i, t_0)$ perpendicular to $\overline{v_i \mathbf{o}}$ (generically the minor axis). The assumption $\mathcal{A}(v_i)$ makes this operation well defined. Further, T_1 is probability preserving since rigid reflection preserves Lebesgue measure.

Critically, T_1 exchanges the events $\mathcal{A}(v_i) \cap \mathcal{B}(v_i) \cap \{X_i = 1\}$ and $\mathcal{A}(v_i) \cap \mathcal{B}(v_i) \cap \{X_i = 0\}$ since $E(v_i, t_0) - B(v_i, h_0) - B(\mathbf{o}, h_0)$ is symmetric about the boundary of $H(v_i, 1/2)$. As a result, $\mathbf{P}(\{X_i = 1\} \cap \mathcal{A}(v_i) \cap \mathcal{B}(v_i)) = \mathbf{P}(\{X_i = 0\} \cap \mathcal{A}(v_i) \cap \mathcal{B}(v_i)$, and hence $\mathbf{P}(X_i = 1) = 1/2 + O(\mathbf{P}(\mathcal{A}(v_i) \cap \mathcal{B}(v_i)))$. Summation and Lem. 5 give

$$\mathbf{E}[X] = n/2 + o(n). \tag{5}$$

Step 2: Our proof of the second step is similar. Fix $i, j, i \neq j$. We construct maps T_2 and T_3 on our probability space K^n. On the event $\mathcal{D}(v_i, v_j)$ both act as the identity. Otherwise, T_2 maps any point except v_i in $E(v_i, t_0) - B(\mathbf{o}, h_0) - B(v_i, h_0)$ to its reflection across the axis of $E(v_i, t_0)$ perpendicular to $\overline{\mathbf{o} v_i}$. Define T_3 analogously, but with respect to v_j.

The key property now is that the regions $E(v_i, t_0) - B(\mathbf{o}, h_0) - B(v_i, h_0)$ and $E(v_j, t_0) - B(\mathbf{o}, h_0) - B(v_j, h_0)$ are disjoint, by Lem. 4 and the choice of parameters promised in Lem. 5. Hence T_2 and T_3 commute (and thus form an Abelian group under composition), and so

$$\mathcal{D}(v_i, v_j) \cap \{X_i = \epsilon_i \wedge X_j = \epsilon_j\}, \quad \epsilon_i, \epsilon_j \in \{0, 1\},$$

are a single orbit of the action of this group. As T_2 and T_3 preserve probability, the above events are equiprobable. Hence, $\mathbf{E}[X_i X_j] = 1/4 + O(\mathbf{P}(\overline{\mathcal{D}(v_i, v_j)}))$. As $\mathbf{E}[X_i] = 1/2 + O(\mathbf{P}(\overline{\mathcal{A}(v_i) \cap \mathcal{B}(v_i)}))$, we see that $\mathbf{Cov}(X_i, X_j) = \mathbf{E}[X_i X_j] - \mathbf{E}[X_i]\mathbf{E}[X_j] = O(\mathbf{P}(\overline{\mathcal{D}(v_i, v_j)}))$. Thus, summation and Lem. 5 gives

$$\mathbf{Var}(X) = \sum_{i=1}^{n} \mathbf{Var}(X_i) + \sum_{i \neq j} \mathbf{Cov}(X_i, X_j) = o(n^2). \tag{6}$$

Lem. 3, (5), and (6) then finish the proof. $\square$

We are then left with the task of verifying Lem. 5. The following necessary geometric arguments we state with only a proof sketch for each.

Lemma 6. *Let v be chosen uniformly in K. If $t \to 0$, the probability that $E(v,t) \not\subseteq K$ is $O(t^2)$.*

One verifies that the ellipse $E(v, t)$ does not extend much farther than v, and so in particular if v avoids the boundary by a slim margin, the result holds.

Lemma 7. *Under the assumption $t_0, r_0, h_0 \to 0$ we have*

$$\mathbf{P}(\overline{\mathcal{B}(v_i)}) = O(r_0^2) + O((1 - t_0 + O(t_0^3))^{n-1}) + O((n-1)h_0^2).$$

$\mathcal{B}(v_i)$ consists of the requirements that v_i avoid $B(\mathfrak{o}, r_0)$, some point besides v_i is contained in $E(v_i, t_0)$, and no points fall in $B(\mathfrak{o}, h_0) \cup B(v_i, h_0)$, and the above expression is simply the union bound of the relevant failure probabilities.

We are finally in a position to prove Lem. 5.

Proof (Lem. 5). By Lem. 6, $\mathbf{P}(\overline{\mathcal{A}(v_i)}) = O(t_0^2)$. By Lem. 7,

$$\mathbf{P}(\overline{\mathcal{B}(v_i)}) = O(r_0^2) + O((1 - t_0 + O(t_0^3))^{n-1}) + O((n-1)h_0^2).$$

It is clear that $\mathbf{P}(\overline{\mathcal{C}(v_i, v_j)}) = \theta_0$, since this is just the probability that a random uniform angle in $[0, 2\pi]$ is at most θ_0. Note that there is a $c_2 > 0$ such that

$$\frac{c_1 t_0^2}{r_0^5(1 - \cos\frac{\theta_0}{2})} \leq \frac{c_2 t_0^2}{r_0^5 \theta_0^2} \tag{7}$$

using the approximation $\cos x \geq 1 - x^2/2 + O(x^4)$ for x sufficiently small, under the assumption $\theta_0 \to 0$ and $t_0^2 r_0^{-5} \to 0$, say.

Setting $t_0 = \frac{3\ln n}{n}$, we see that the term $O(t_0^2) = O(n^{-2}\ln^2 n)$ and $O((1 - t_0 + O(t_0^3))^{n-1}) = O(n^{-2.99})$. We set $h_0 = c_2 \frac{t_0^2}{r_0^5 \theta_0^2}$ as in (7), thus making h_0 a threshold as in Lem. 4. Taking $\theta_0 = n^{-3/10}$ and $r_0 = n^{-3/20}$, yields the bound

$$\mathbf{P}(\overline{\mathcal{D}(v_i, v_j)}) = O(n^{-3/10}\ln^2 n).$$

Since $\mathbf{P}(\overline{\mathcal{A}(v_i) \cap \mathcal{B}(v_i)}) \leq \mathbf{P}(\overline{\mathcal{D}(v_i, v_j)}))$, the sums claimed in the lemma follow. $\square$

Remark 4. The proof for $\mathcal{T}_\alpha$ requires a slightly more careful approach. In particular, one needs to account for the fact that the γ-region about v_i in which one expects to find another point becomes increasingly slender as $i \to n$, while at the same time the likelihood of finding a point near o increases. We can partition the vertices $v_1, \ldots, v_n$ into epochs, where vertex i is in epoch $\lfloor \lg i \rfloor$. For each such set, the appropriate value of t_0, r_0, θ_0 etc. will change. The challenge then becomes analyzing cross-terms $\mathbf{E}[X_i X_j]$ when i and j lie in different sets, which requires slightly refined geometric estimates and significantly more bookkeeping.

Proof (Remainder of proof of Thm. 5). With β fixed, we shall assume in what follows that $|\alpha - 1| = o(\gamma)$ for a parameter $\gamma \to 0$. Further, we assume that p is a point such that $\|o - p\| = r$ is bounded away from zero.

We consider the γ-region about p, and construct inscribed and circumscribed figures about the γ-region. Let l_1 be the intersection of the line through o and p and the γ region. The length of l_1 is thus $(1 + o(1))r$. Let l_2 be the segment perpendicular to l_1 which intersects the midpoint of $\overline{po}$. The length of l_2 is $(1+o(1))r\sqrt{\gamma}$. The convex hull of $l_1 \cup l_2$ forms a rhombus R_1 contained entirely in the γ-influence region with area $(1+o(1))r^2\sqrt{\gamma}$. We can also form a circumscribed rectangle R_2 which is axis-parallel to the segments l_1 and l_2, and which has asymptotically twice the area of the rhombus.

Now, let p be one of our randomly chosen points v_i, and further assume it has distance $\|v_i - o\|$ bounded away from 0 and $1/\sqrt{\pi}$, which happens with positive probability. Next, set $\gamma = (nr)^{-2}$, which causes the rhombus and rectangle to have areas $(1 + o(1))n^{-1}$ and $(1 + o(1))2n^{-1}$, respectively. Further, for n sufficiently large the rhombus, ellipse, and rectangle all lie in K. Thus, the event that one of $\{v_1, \ldots, v_n\} - \{v_i\}$ falls in the rhombus and all others avoid the rectangle is at least

$$\binom{n}{1} \text{Area}(R_1)(1 - \text{Area}(R_2))^{n-2} = \binom{n}{1} n^{-1}(1+o(1))(1 - 2n^{-1}(1+o(1)))^{n-2} > c_1 > 0,$$

for some positive constant c_1.

Now, for γ bounded above and hence the length of l_2 bounded above, we see that

$$\frac{\text{Area}(H(v_i, \beta) \cap R_1)}{\text{Area}(R_1)} > c_2 > 0.$$

Summarizing, the probability that there is exactly one point in the rhombus apart from v_i and that all other points fall outside the rectangle is bounded below by some positive constant. Further, the probability that a point chosen uniformly in the rhombus falls in $H(v_i, \beta)$ is bounded below by a positive constant. Thus, the total probability that the target of v_i is local is bounded below by a positive constant, which shows that $\mathbf{E}[X_i] > c > 0$ for some constant c, hence $\mathbf{E}[\rho(\beta)] > c > 0$. The upper bound is similar.

Acknowledgments

The author wishes to thank Fan Chung for starting him on this project and for all her assistance. He would additionally like to thank Bill Chen and the Center for Combinatorics at Nankai University for their hospitality during part of this research.

References

1. Fabrikant, A., Koutsoupias, E., Papadimitriou, C.: Heuristically optimized trade-offs: A new paradigm for powerlaws in the internet. In: Proc. of 29th International Colloquium of Automata, Languages, and Programming (2002)
2. Bollobás, B., Chung, F.: The diameter of a cycle plus a random matching. SIAM J. on Discrete Math, 328–333 (1988)
3. Kleinberg, J.: The small-world phenomenon: An algorithmic perspective. In: Proc. 32nd ACM Symposium on Theory of Computing (2000)
4. Liben-Nowell, D., Novak, J., Kumar, R., Raghavan, P., Tomkins, A.: Geographic routing in social networks. Proceedings of the National Academy of Sciences 102(33), 11623–11628 (2005)
5. Andersen, R., Chung, F., Lu, L.: Modeling the small-world phenomena with local network flow. Internet Mathematics, 359–385 (2006)
6. Chung, F., Lu, L.: Complex Graphs and Networks. CBMS Lecture Notes. AMS (2006)
7. Castells, M.: The Internet Galaxy: Reflections on the Internet, Business, and Society. Oxford U. Press, New York (2001)
8. Berger, N., Bollobás, B., Borgs, C., Chayes, J., Riordan, O.:Degree distribution of the FKP network model. In: Proc. of the 30th International Colloquium of Automata, Languages and Programming, pp. 725–738 (2003)
9. Penrose, M.: Geometric Random Graphs. Oxford U. Press, New York (2003)
10. Meester, R., Roy, R.: Continuum Percolation. Cambridge U. Press, New York (1996)
11. Flaxman, A., Frieze, A., Vera, J.: Geometric preferential attachment model of networks. In: Proc. 3rd International Workshop on Algorithms and Models for the Web-Graph, pp. 44–55 (2004)
12. Chung, F., Jungreis, D., Handjani, S.: Generalizations of Pólya's urn problem. Ann. of Comb, 141–153 (2003)
13. Alon, N., Spencer, J.: The Probabilistic Method. John Wiley & Sons Inc, Chichester (2000)

Approximating PageRank from In-Degree

Santo Fortunato[1,2], Marián Boguñá[3],
Alessandro Flammini[1], and Filippo Menczer[1]

[1] School of Informatics, Indiana University
Bloomington, IN 47406, USA
[2] Complex Networks Lagrange Laboratory (CNLL),
ISI Foundation, Torino, Italy
[3] Departament de Física Fonamental, Universitat de Barcelona
08028 Barcelona, Spain

Abstract. PageRank is a key element in the success of search engines, allowing to rank the most important hits in the top screen of results. One key aspect that distinguishes PageRank from other prestige measures such as in-degree is its global nature. From the information provider perspective, this makes it difficult or impossible to predict how their pages will be ranked. Consequently a market has emerged for the optimization of search engine results. Here we study the accuracy with which PageRank can be approximated by in-degree, a local measure made freely available by search engines. Theoretical and empirical analyses lead to conclude that given the weak degree correlations in the Web link graph, the approximation can be relatively accurate, giving service and information providers an effective new marketing tool.

1 Introduction

PageRank has become a key element in the success of Web search engines, allowing to rank the most important hits in the top page of results. Certainly the introduction of PageRank as a factor in sorting results [1] has contributed considerably to Google's lasting dominance in the search engine market [2].

PageRank is not the only possible measure of importance or prestige among Web pages. The simplest possible way to measure the prestige of a page is to count the incoming links (in-links) to the page. There is a correlation between the number of in-links that a page receives from other pages (in-degree) and quality, especially when the in-degree is large. The in-degree of Web pages is very cheap to compute and maintain, so that a search engine can easily keep in-degree updated with the evolution of the Web.

However, in-degree is a local measure. All links to a page are considered equal, regardless of where they come from. Two pages with the same in-degree are considered equally important, even if one is cited by more prestigious sources than the other. To modulate the prestige of a page with that of the pages pointing to it means to move from the examination of an individual node in the link graph to that of the node together with its predecessor neighbors. PageRank represents

W. Aiello et al. (Eds.): WAW 2006, LNCS 4936, pp. 59–71, 2008.

such a shift from the local measure given by in-degree toward a global measure where each Web page contributes to define the importance of every other page.

The use of PageRank in place of in-degree for applications such as ranking by Web search engines relies on two assumptions: (i) PageRank is a quantitatively different and better prestige measure compared to in-degree; and (ii) PageRank cannot be easily guessed or approximated by in-degree. To wit, Amento et al. [3] report a very high average correlation between in-degree and PageRank (Spearman $\rho = 0.93$, Kendall $\tau = 0.83$) based on five queries. Further, they report the same average precision at 10 (60%) based on relevance assessments by human subjects. In this paper we further quantitatively explore these assumptions answering the following questions: *What is the correlation between in-degree and PageRank across representative samples of the Web? How accurately can one approximate PageRank from local knowledge of in-degree?*

From the definition of PageRank, other things being equal, the PageRank of a page grows with the in-degree of the page. Beyond this zero-order approximation, the actual relation between PageRank and in-degree has not been thoroughly investigated in the past. It is known that the distributions of PageRank and in-degree follow an almost identical pattern [4,5], i.e., a curve ending with a broad tail that follows a power law with exponent $\gamma \simeq 2.1$. This fact may indicate a strong correlation between the two variables. Surprisingly there is no agreement in prior literature about the correlation between PageRank and in-degree. Pandurangan et al. [4] show very little correlation based on analysis of the brown.edu domain and the TREC WT10g collection. Donato et al. [5] report on a correlation coefficient which is basically zero based on analysis of a much larger sample ($2 \cdot 10^8$ pages) taken from the WebBase [6] collaboration. On the other hand, analysis of the University of Notre Dame domain by Nakamura [7] reveals a strong correlation.

In Section 2 we estimate PageRank for a generic directed network within a mean field approach. For a network without degree-degree correlations the average PageRank turns out to be simply proportional to the in-degree, modulo an additive constant. The prediction is validated empirically in Section 3, where we solve the equations numerically for four large samples of the Web graph; in each case the agreement between our theoretical estimate and the empirical data is excellent. We find that the Web graph is basically uncorrelated, so the average PageRank can be well approximated by a linear function of the in-degree. As an additional contribution we settle the issue of the correlation between PageRank and in-degree; the linear correlation coefficient is consistently large for all four samples we have examined, in agreement with Nakamura [7]. Finally, in Section 4, we present an application of our findings on the live Web.

2   Theoretical Analysis

The PageRank $p(i)$ of a page i is defined through the following expression [1]:

$$p(i) = \frac{q}{N} + (1 - q) \sum_{j:j \to i} p(j)/k_{out}(j) \qquad i = 1, 2, \ldots, N \qquad (1)$$

where N is the total number of pages, $j \to i$ indicates a hyperlink from j to i, $k_{out}(j)$ is the out-degree of page j and q is the so-called teleportation (or jumping) factor. The set of Equations 1 can be solved iteratively. From Eq. 1 it is clear that the PageRank of a page grows with the PageRank of the pages that point to it. However, the sum over predecessor neighbors implies that PageRank also increases with the in-degree of the page.

PageRank can be thought of as the stationary probability of a random walk process with additional random jumps. The physical description of the process is as follows: when a random walker is in a node of the network, at the next time step with probability q it jumps to a randomly chosen node and with probability $1 - q$ it moves to one of its successors with uniform probability. In the case of directed networks, a node may have no successors. In this case the walker jumps to a randomly chosen node of the network with probability one. The PageRank of a node i, $p(i)$, is then the probability to find the walker at node i when the process has reached the steady state, a condition that is always guaranteed by the teleportation probability q.

The probability to find the walker at node i at time step n follows a simple Markovian equation:

$$p_n(i) = \frac{q}{N} + (1 - q) \sum_{j:k_{out}(j) \neq 0} \frac{a_{ji}}{k_{out}(j)} p_{n-1}(j) + \frac{1-q}{N} \sum_{j:k_{out}(j)=0} p_{n-1}(j), \quad (2)$$

where a_{ji} is the adjacency matrix with entry 1 if there is a direct connection between j and i and zero otherwise. The first term in Eq. 2 is the contribution of walkers jumping to a randomly chosen node, the second term is the random walk contribution, and the third term accounts for walkers that at the previous step were located in dangling nodes and now jump to random nodes. In the limit $n \to \infty$ this last contribution becomes a constant term affecting all the nodes in the same way, and thus it can be removed from Eq. 2 under the constraint that the final solution is properly normalized. Strictly speaking this would lead to an effective teleportation term, which we omit to keep the notation simple. Alternatively dangling nodes could be taken into account by a proper rescaling of the the the second term [8]. Hereafter we intend all sums over nodes to exclude dangling ends, considering only nodes with $k_{out} > 0$. The PageRank of page i is the steady state solution of Eq. 2, $p(i) = \lim_{n \to \infty} p_n(i)$. Equation 2 cannot be analytical solved. We propose a mean field solution of Eq. 2 that, nevertheless, gives a very accurate description of the PageRank structure of the Web. The mean field approach is often used in statistical physics, and is reliable when each element of the system has many interaction partners,[1] as in this case the effect of the interactions can be taken into account in an average way, neglecting the variations among the elements.

Instead of analyzing the PageRank of single pages, we aggregate pages in classes according to their degree $\mathbf{k} \equiv (k_{in}, k_{out})$ and define the average PageRank of nodes of degree class $\mathbf{k}$ as

---

[1] On hypercubic lattices, the mean field limit for most spin models is reached in four dimensions, when each spin has eight neighbors.

$$\bar{p}_n(\mathbf{k}) \equiv \frac{1}{NP(\mathbf{k})} \sum_{i \in \mathbf{k}} p_n(i). \tag{3}$$

Note that now "degree class $\mathbf{k}$" means all the nodes with in-degree k_{in} and out-degree k_{out}; $P(\mathbf{k})$ is the probability that a node is in the degree class $\mathbf{k}$. Taking the average of Eq. 2 for all nodes of the degree class $\mathbf{k}$ we obtain

$$\frac{1}{NP(\mathbf{k})} \sum_{i \in \mathbf{k}} p_n(i) = \frac{q}{N} + \frac{(1-q)}{NP(\mathbf{k})} \sum_{i \in \mathbf{k}} \sum_{j:k_{out}(j) \neq 0} \frac{a_{ji}}{k_{out}(j)} p_{n-1}(j). \tag{4}$$

From Eq. 3 we see that the left-hand side of Eq. 4 is $\bar{p}_n(\mathbf{k})$. In the right-hand side we split the sum over j into two sums, one over all the degree classes $\mathbf{k}'$ and the other over all the nodes within each degree class $\mathbf{k}'$. We get

$$\bar{p}_n(\mathbf{k}) = \frac{q}{N} + \frac{(1-q)}{NP(\mathbf{k})} \sum_{\mathbf{k}'} \frac{1}{k'_{out}} \sum_{i \in \mathbf{k}} \sum_{j \in \mathbf{k}'} a_{ji} p_{n-1}(j). \tag{5}$$

At this point we perform our mean field approximation [9], which consists in substituting the PageRank of the predecessor neighbors of node i by its mean value, that is,

$$\sum_{i \in \mathbf{k}} \sum_{j \in \mathbf{k}'} a_{ji} p_{n-1}(j) \simeq \bar{p}_{n-1}(\mathbf{k}') \sum_{i \in \mathbf{k}} \sum_{j \in \mathbf{k}'} a_{ji}$$

$$= \bar{p}_{n-1}(\mathbf{k}') E_{\mathbf{k}' \to \mathbf{k}}, \tag{6}$$

where $E_{\mathbf{k}' \to \mathbf{k}}$ is the total number of links pointing from nodes of degree $\mathbf{k}'$ to nodes of degree $\mathbf{k}$. This matrix can also be rewritten as

$$E_{\mathbf{k}' \to \mathbf{k}} = k_{in} P(\mathbf{k}) N \frac{E_{\mathbf{k}' \to \mathbf{k}}}{k_{in} P(\mathbf{k}) N}$$

$$= k_{in} P(\mathbf{k}) N P_{in}(\mathbf{k}'|\mathbf{k}), \tag{7}$$

where $P_{in}(\mathbf{k}'|\mathbf{k})$ is the probability that a predecessor of a node belonging to degree class $\mathbf{k}$ belongs to degree class $\mathbf{k}'$. The conditional probability $P_{in}(\mathbf{k}'|\mathbf{k})$ incorporates the so-called *degree-degree correlation*, i.e., the correlation between the degree of a node and that of its neighbors (see [10] pp. 243–245). Using Equations 6 and 7 in Eq. 5 we finally obtain

$$\bar{p}_n(\mathbf{k}) = \frac{q}{N} + (1-q) k_{in} \sum_{\mathbf{k}'} \frac{P_{in}(\mathbf{k}'|\mathbf{k})}{k'_{out}} \bar{p}_{n-1}(\mathbf{k}'), \tag{8}$$

which is a closed set of equations for the average PageRank of pages in the same degree class. When the network has degree-degree correlations, the solution of this equation is non-trivial and the resulting PageRank can have a complex dependence on the degree. However, in the particular case of networks without degree-degree correlations, the transition probability $P_{in}(\mathbf{k'}|\mathbf{k})$ becomes independent of $\mathbf{k}$ and takes the simpler form

$$P_{in}(\mathbf{k'}|\mathbf{k}) = \frac{k'_{out}P(\mathbf{k'})}{\langle k_{in}\rangle}, \tag{9}$$

where $\langle\cdot\rangle$ denotes the average value of the quantity in brackets. Using this expression in Eq. (8) and taking the limit $n \to \infty$, we obtain

$$\overline{p}(\mathbf{k}) = \frac{q}{N} + \frac{1-q}{N}\frac{k_{in}}{\langle k_{in}\rangle}, \tag{10}$$

that is, the average PageRank of nodes of degree class $\mathbf{k}$ is independent of k_{out} and proportional to k_{in}.

The same type of analysis allows to estimate the size of the fluctuations of PageRank for nodes in the same degree class $\mathbf{k}$. It turns out that, for uncorrelated networks, the standard deviation $\sigma(\mathbf{k})$ of the PageRank distribution about its mean value is

$$\sigma^2(\mathbf{k}) \simeq \frac{(1-q)^4}{N^2\langle k_{in}\rangle^3}\left\langle\frac{k_{in}^2}{k_{out}}\right\rangle k_{in}. \tag{11}$$

For large in-degrees, the coefficient of variation is

$$\frac{\sigma(\mathbf{k})}{\overline{p}(\mathbf{k})} \simeq (1-q)\left[\left\langle\frac{k_{in}^2}{k_{out}}\right\rangle\frac{1}{\langle k_{in}\rangle k_{in}}\right]^{1/2}. \tag{12}$$

The factor $\left\langle\frac{k_{in}^2}{k_{out}}\right\rangle$ in this expression can be very large when the network has a long-tailed degree distribution, which implies that the relative fluctuations are large for small in-degrees. Therefore the true PageRank of pages with small in-degree may differ significantly from its mean field approximation. However, for large in-degrees the relative fluctuations become less important — due to the factor k_{in} in the denominator — and the average PageRank from Eq. 10 gives a good approximation. Note that the expression in Eq. 12 relates to the relative fluctuations *within* a degree class, rather than across the entire graph. Since PageRank is distributed according to a power law with γ close to 2, the overall fluctuations diverge in the limit of infinite graph size. An analysis of the PageRank distribution and of the relative fluctuations within each degree class is omitted here for brevity, and will be included in an extended version of this paper.

3 Results

For an empirical validation of the theoretical predictions in the previous section, we analyzed four samples of the Web graph. Two of them were obtained by crawls performed in 2001 and 2003 by the WebBase collaboration [6]. The other two were collected by the WebGraph project [11]: the pages belong to two national

Table 1. Number of pages, links, and average degree ($\langle k \rangle = \langle k_{in} \rangle = \langle k_{out} \rangle$) for the four data sets we have analyzed

Data set	WB 2001	.uk 2002	WB 2003	.it 2004
# pages	8.1×10^7	1.9×10^7	4.9×10^7	4.1×10^7
# links	7.5×10^8	2.9×10^8	1.2×10^9	1.1×10^9
$\langle k \rangle$	9.34	15.78	24.05	27.50

Table 2. Exponents of the power law part of the PageRank distribution and linear correlation coefficients between PageRank and in-degree

Data set	WB 2001	.uk 2002	WB 2003	.it 2004
γ	2.2 ± 0.1	2.0 ± 0.1	2.0 ± 0.1	2.0 ± 0.1
ρ	0.538	0.554	0.483	0.733

domains, .uk (2002) and .it (2004), respectively. In Table 1 we list the number of vertices and edges and the average degree for each data set.

We calculated PageRank with the standard iterative procedure; the factor q was set to 0.15, as in the original paper by Brin and Page [1] and many successive studies. In Fig. 1 we show the cumulative distributions of PageRank, i.e. the function $R(p)$ representing the probability that PageRank exceeds the value p. Using the cumulative distribution allows to reduce the noise due to fluctuations at large PageRank values. In all four cases we obtained a pattern with a broad tail. The initial part of the distribution can be well fitted by a power law $p^{-\beta}$ with exponent β between 1.0 and 1.2. The exponents for the actual PageRank distribution are $\gamma = \beta + 1$, so they range from 2.0 to 2.2, in agreement with other studies [4,5]. The right-most part of each curve, corresponding to the pages with highest PageRank, decreases faster. For the WebBase sample of 2001 the tail of the curve up to the last point can be well fitted by a power law with exponent $\beta \approx 1.6$; in the other cases we see evidence of an exponential cutoff.

We also calculated the linear correlation coefficient between PageRank and in-degree. In Table 2 we list Pearson's ρ together with the slope of the power law portions of the PageRank distributions. The correlation between PageRank and in-degree is rather strong, in contrast to the findings of [4] and especially [5] but in agreement with [7] and consistently with the high correlation observed between in-degree and Kleinberg's authority score [12].

Let us now validate the expression derived from our mean field analysis for the average PageRank. We solved Eq. 8 with an analogous iterative procedure as the one we used to calculate PageRank. We now look for the vector $\bar{p}(\mathbf{k})$, defined for all pairs $\mathbf{k} \equiv (k_{in}, k_{out})$ which occur in the network. Since PageRank is a probability, it must be normalized so that its sum over all vertices of the network is one. So we initialized the vector with the constant $\bar{p}_0(\mathbf{k}) = 1/N$, and plugged it into the right-hand side of Eq. 8 to get the first approximation $\bar{p}_1(\mathbf{k})$. We then used $\bar{p}_1(\mathbf{k})$ as input to get $\bar{p}_2(\mathbf{k})$, and so on. We remark that the expression of the probability $P_{in}(\mathbf{k}'|\mathbf{k})$ is not a necessary ingredient of the

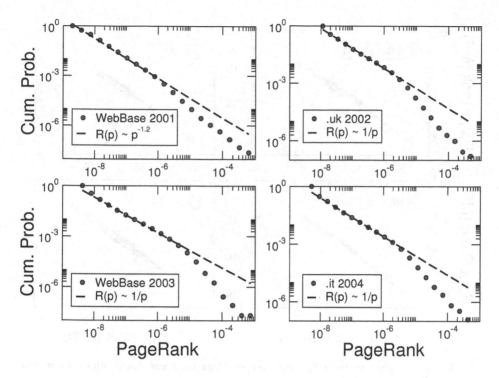

Fig. 1. Cumulative distributions of PageRank

calculation. In fact, the sum on the right-hand side of Eq. 8 is just the average value of $\bar{p}_{n-1}(\mathbf{k}')/k'_{out}$ among all predecessors of vertices with degree $\mathbf{k}$. The algorithm leads to convergence within a few iterations (we never needed more than 20). In Fig. 2 we compare the values of $\bar{p}(\mathbf{k})$ calculated from Eq. 8 with the corresponding empirical values. Here we averaged $\bar{p}(\mathbf{k})$ over out-degree, so it only depends on the in-degree k_{in}. The variation of $\bar{p}(\mathbf{k})$ with k_{out} (for fixed k_{in}) turns out to be very small. The scatter plots of Fig. 2 show that the mean field approximation gives excellent results: the points are very tightly concentrated about each frame bisector, drawn as a guide to the eye.

Next let us analyze explicitly the relation between PageRank and in-degree. To plot the function $\bar{p}(k_{in})$ directly is not very helpful because the wide fluctuations of PageRank within each degree class would mystify the pattern for large values of k_{in}. So we average PageRank within bins of in-degree, which is the standard procedure to derive trends from scatter plots (see [10] pp. 240–242). As both PageRank and in-degree are power-law distributed, we use logarithmic bins; the multiplicative factor for the bin size is 1.3. The resulting patterns for our four Web samples are presented in Fig. 3. The empirical curves are rather smooth, and show that the average PageRank (per degree class) is an increasing function of in-degree. The relation between the two variables is approximately linear

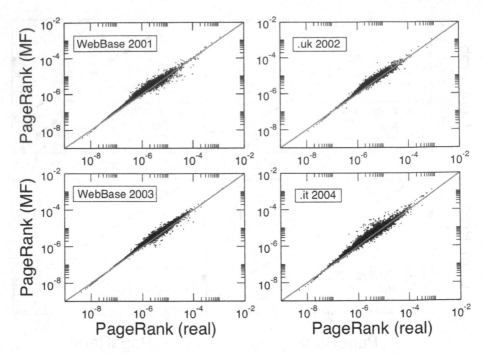

Fig. 2. Scatter plots of the empirical average PageRank per degree class versus our mean field (MF) estimate

for large in-degrees. This is exactly what we would expect if the degrees of pages were uncorrelated with those of their neighbors in the Web graph (cf. Section 2). In such a case the relation between PageRank and in-degree is given by Eq. 10. Indeed, the comparison of the empirical data with the curves of Eq. 10 in Fig. 3 is quite good for all data sets. We infer that the Web graph is an essentially uncorrelated graph; this is confirmed by direct measurements of degree-degree correlations in our four Web samples [13]. What is most important, the average PageRank of a page with in-degree k_{in} is well approximated by the simple expression of Eq. 10.

4   Applications to the Live Web

Knowing the relationship between PageRank and in-degree has potential applications for the Web graph. It is vital for many service and information providers to have good rankings by major search engines for relevant keywords, given that search engines are the primary way that Internet users find and visit Web sites [14,15]. Consequently a demand has emerged for companies that perform so-called *search engine optimization* or *search engine marketing* on behalf of business clients. The goal is to increase the rankings of their pages, thus directing traffic to their sites [16].

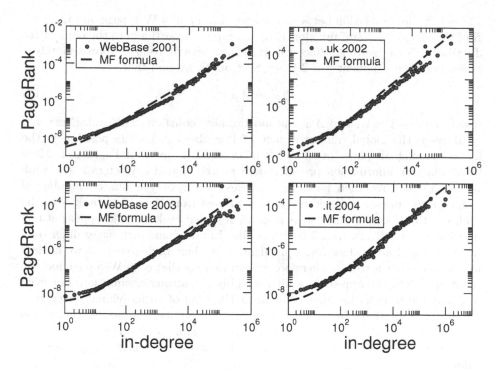

Fig. 3. PageRank versus in-degree; the dashed line is the approximation given by the closed formula of Eq. 10

In the previous section we have shown that the average PageRank of a page with in-degree k_{in} can be well approximated by the closed formula in Eq. 10. So Web authors may use local in-degree information as a proxy for estimating the global PageRank of their sites.

To use Eq. 10 for the Web we need to know the total number N of Web pages indexed by a search engine, say Google, and their average degree $\langle k_{in} \rangle$. The size of the Google index was published until recently; we use the last reported number, $N \simeq 8.1 \times 10^9$. The average degree is not known; the best we can do is extract it from samples of the Web graph. Our data sets do not deliver a unique value for $\langle k_{in} \rangle$, but they agree on the order of magnitude (see Table 1). Hereafter we use $\langle k_{in} \rangle = 10$.

Let us now consider whether Eq. 10 can be useful in the live Web. Ideally we should compare the PageRank values of a list of Web pages with the corresponding values derived through our formula. Unfortunately the real PageRank values calculated by a search engine such as Google are not accessible, so we need a different strategy. The simplest choice is to focus on rank rather than PageRank. We know that Google ranks Web pages according to their PageRank values as well as other features which do not depend on Web topology. The latter features are not disclosed; in the following we disregard them and assume for simplicity that the ranking of a Web page exclusively depends on its PageRank value.

There is a simple relation between the PageRank p of a Web page and the rank R of that page. The Zipf function $R(p)$ is simply proportional to the cumulative distribution of PageRank. Since the PageRank distribution is approximately a power law with exponent $\gamma \simeq 2.1$ (see Section 3), we find that

$$R(p) \simeq A p^{-\beta}, \tag{13}$$

where $\beta = \gamma - 1 \simeq 1.1$ and A is a proportionality constant. The rank R referred to above is the global rank of a page of PageRank p, i.e., its position in the list containing all pages of the Web in decreasing order of PageRank. More interesting for information providers and search engine marketers is the rank within hit lists returned for actual queries, where only a limited number of result pages appear. We need a criterion to pass from the global rank R to the rank r within a query's hit list. A page with global rank R could appear at any position $r = 1, 2, \ldots, n$ in a list with n hits. In our framework pages differ only by their PageRank values (or, equivalently, by their in-degrees), as we neglect lexical and other features. Therefore we can assume that each Web page has the same probability to appear in a hit list. This is a strong assumption, but even if it may fail to describe what happens at the level of an individual query, it is a fair approximation when one considers a large number of queries. Under this hypothesis the probability distribution of the possible positions is a Poissonian, and the expected local rank r of a page with global rank R is given by the mean value:

$$r = R \frac{n}{N}. \tag{14}$$

Now it is possible to test the applicability of Eq. 10 to the Web. We are able to estimate the rank of a Web page within a hit list if we know the number of in-links k_{in} of the page and the number n of hits in the list. The procedure consists of three simple steps:

1. from k_{in} we calculate the PageRank p of the page according to Eq. 10;
2. from p we determine the global rank R according to Eq. 13;
3. from R and n we derive the local rank r according to Eq. 14.

The combination of the three steps leads to the following expression of the local rank r as a function of k_{in} and n:

$$r = \frac{A n}{(\frac{q}{N} + \frac{1-q}{N\langle k_{in} \rangle} k_{in})^{1.1} N}. \tag{15}$$

We remark that A is a simple multiplicative constant, and its value has no effect on the dependence of the local rank r on the variables k_{in} and n. Therefore we decided to consider it as a free parameter, whose value is to be determined by the comparison with empirical data.

For our analysis we used a set of $65,207$ actual queries from a September 2001 AltaVista log. We submitted each query to Google, and picked at random one of the pages of the corresponding hit list. For each selected page, we stored its actual rank r_{emp} within the hit list, as well as its number k_{in} of in-links,

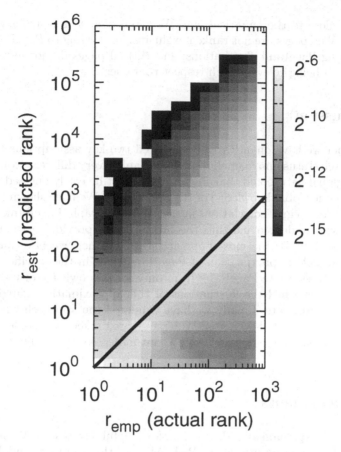

Fig. 4. Density map of the scatter plot between predicted rank r_{est} and actual rank r_{emp} for 65,207 queries. The fraction of points in each log-size bin is expressed by the color, also on a logarithmic scale. The diagonal guide to the eye is $r_{est} = r_{emp}$.

which was again determined through Google.[2] The number n of hits of the list was also stored. Google (like other search engines) never displays more than 1000 results per query, so we always have $r_{emp} \leq 1000$. From k_{in} and n we estimated the theoretical rank r_{est} by means of Eq. 15, and compared it with its empirical counterpart r_{emp}. The comparison can be seen in the scatter plot of Fig. 4. Given the large number of queries and the broad range of rank values, we visualize the density of points in logarithmic bins. The region with highest density is a stripe centered on the diagonal line $r_{est} = r_{emp}$ by a suitable choice of A ($A = 1.5 \times 10^{-4}$). We conclude that the rank derived through Eq. 15 is in

---

[2] The in-degree data provided by search engines is only an estimate of the true number. First, a search engine can only know of links from pages that it has crawled and indexed. Second, for performance reasons, the algorithms counting inlinks use various unpublished approximations based on sampling.

most cases close to the empirical one. We stress that this result is not trivial, because (i) Web pages are not ranked exclusively according to PageRank; (ii) we are neglecting PageRank fluctuations; and (iii) all pages do not have the same probability of being relevant with respect to a query.

5 Discussion

In this paper we have quantitatively explored two key assumptions around the current search status quo, namely that PageRank is very different from in-degree due to its global nature and that PageRank cannot be easily guessed or approximated without global knowledge of the Web graph. We have shown that due to the weak degree-degree correlations in the Web link graph, PageRank is strongly correlated with in-degree and thus the two measures provide very similar information, especially for the most popular pages. Further, we have introduced a general mean field approximation of PageRank that, in the specific case of the Web, allows to estimate PageRank from only local knowledge of in-degree. We have further quantified the fluctuations of this approximation, gauging the reliability of the estimate. Finally we have validated the approach with a simple procedure that predicts how actual Web pages are ranked by Google in response to actual queries, using only knowledge about in-degree and the number of query results.

Acknowledgments

We thank A. Vespignani and M. Serrano, for helpful discussions. We are grateful to Google for extensive use of its Web API, to the WebBase and WebGraph projects for their crawl data, and to AltaVista for use of their query logs. This work is funded in part by a Volkswagen Foundation grant to SF, by the Spanish government's DGES grant FIS2004-05923-CO2-02 and Generalitat de Catalunya grant SGR00889 to MB, by NSF Career award 0348940 to FM, and by the Indiana University School of Informatics.

References

1. Brin, S., Page, L.: The anatomy of a large-scale hypertextual Web search engine. Computer Networks 30(1–7), 107–117 (1998)
2. Sullivan, D.: Nielsen//netratings search engine ratings (August (2005), http://searchenginewatch.com/reports/article.php/2156451
3. Amento, B., Terveen, L., Hill, W.: Does "authority" mean quality? Predicting expert quality ratings of Web documents. In: Proc. 23rd ACM SIGIR Conf. on Research and Development in Information Retrieval, pp. 296–303 (2000)
4. Pandurangan, G., Raghavan, P., Upfal, E.: Using pagerank to characterize Web structure. In: H. Ibarra, O., Zhang, L. (eds.) COCOON 2002. LNCS, vol. 2387, pp. 330–339. Springer, Heidelberg (2002)

5. Donato, D., Laura, L., Leonardi, S., Millozzi, S.: Large scale properties of the webgraph. European Physical Journal B 38, 239–243 (2004)
6. Garcia-Molina, H.: The Stanford WebBase Project (2005),
   http://www-diglib.stanford.edu/~testbed/doc2/WebBase/
7. Nakamura, I.: Large scale properties of the webgraph. Physical Review 68 (2003) 045104
8. Volkovich, Y., Litvak, N., Donato, D.: Determining factors behind the PageRank log-log plot. Technical Report 1823, Department of Applied Mathematics, University of Twente (2007)
9. Binney, J., Dowrick, N., Fisher, A., Newman, M.: The theory of critical phenomena. First edn. Oxford University Press, Oxford (1992)
10. Pastor-Satorras, R., Vespignani, A.: Evolution and Structure of the Internet. Cambridge University Press, Cambridge, UK (2004)
11. Laboratory for Web Algorithmics (LAW), University of Milan: WebGraph (2005),
    http://webgraph.dsi.unimi.it
12. Donato, D., Leonardi, S., Tsaparas, P.: Stability and similarity of link analysis ranking algorithms. In: Caires, L., Italiano, G.F., Monteiro, L., Palamidessi, C., Yung, M. (eds.) ICALP 2005. LNCS, vol. 3580, pp. 717–729. Springer, Heidelberg (2005)
13. Serrano, M., Maguitman, A., Boguñá, M., Fortunato, S., Vespignani, A.: Decoding the structure of the WWW: Facts versus bias. In: ACM Transactions on the Web (In press)
14. Websidestory: User navigation behavior to effect link popularity (May, Cited by Search Engine Round Table According to this source, Websidestory Vice President Jay McCarthy announced at the Search Engine Strategies Conference (Toronto 2005) that the number of page referrals from search engines has surpassed those from other pages (2005), http://www.seroundtable.com/archives/001901.html
15. Qiu, F., Liu, Z., Cho, J.: Analysis of user web traffic with a focus on search activities. In: Proc. International Workshop on the Web and Databases (WebDB). (2005)
16. Sullivan, D.: Intro to search engine optimization,
    http://searchenginewatch.com/webmasters/article.php/2167921

Probabilistic Relation between In-Degree and PageRank

Nelly Litvak, Werner R.W. Scheinhardt, and Yana Volkovich

University of Twente, Dept. of Applied Mathematics,
P.O. Box 217, 7500AE Enschede, The Netherlands
{n.litvak,w.r.w.scheinhardt,y.volkovich}@ewi.utwente.nl

Abstract. This paper presents a novel stochastic model that explains the relation between power laws of In-Degree and PageRank. PageRank is a popularity measure designed by Google to rank Web pages. We model the relation between PageRank and In-Degree through a stochastic equation, which is inspired by the original definition of PageRank. Using the theory of regular variation and Tauberian theorems, we prove that the tail distributions of PageRank and In-Degree differ only by a multiplicative constant, for which we derive a closed-form expression. Our analytical results are in good agreement with Web data.

Categories and Subject Descriptors
H.3.3:[**Information Storage and Retrieval**]: Information Search and Retrieval– *Retrieval models*; G.3:[**Mathematics of Computing**]: Probability and statistics – *Stochastic processes, Distribution functions*

General Terms
Theory, Verification, Experimentation, Algorithms

Keywords: PageRank, In-Degree, Power law, Regular variation, Stochastic equation, Web measurement.

1 Introduction

We study the relation between the probability distributions of the PageRank and the In-Degree of a randomly selected Web page. In this paper we present the mathematical model and main results while more detailed discussion and proofs can be found in the extended version [1]. The notion of *PageRank* was introduced by Google in order to numerically characterize the popularity of Web pages. The original description of PageRank presented in [2] is as follows:

$$PR(i) = c \sum_{j \to i} \frac{1}{d_j} PR(j) + (1 - c), \tag{1}$$

where $PR(i)$ is the PageRank of page i, d_j is the number of outgoing links of page j, the sum is taken over all pages j that link to page i, and c is the "damping factor", which is some constant between 0 and 1. The *In-Degree* of a Web page denotes simply the number of incoming hyperlinks to that page. From

W. Aiello et al. (Eds.): WAW 2006, LNCS 4936, pp. 72–83, 2008.
© Springer-Verlag Berlin Heidelberg 2008

equation (1) it is clear that the PageRank of a page depends on its In-Degree and the importance (i.e. PageRanks) of the pages that link to it.

We focus in particular on the *tail asymptotics* for PageRank and its connection to In-Degree. By tail of the PageRank distribution we simply mean the fraction of pages $\mathbb{P}(PR > x)$ having PageRank greater than x, where x is large. A common way to analyze tail behavior is to find an asymptotic expression $p(x)$ such that $\mathbb{P}(PR > x)/p(x) \to 1$ as $x \to \infty$. In this case, $p(x)$ and $\mathbb{P}(PR > x)$ are asymptotically equivalent, and thus, we can approximate $\mathbb{P}(PR > x)$ by $p(x)$ for large enough x.

Pandurangan et al. [3] observed that the tails of PageRank and In-Degree distributions for Web data seem to follow power laws with the same exponent. Recent extensive experiments by Donato et al. [4] and Fortunato et al. [5] confirmed this phenomenon. Becchetti and Castillo [6] investigated the influence of the damping factor c on the power law behavior of PageRank. They have shown that the PageRank of the top 10% of the nodes always follows a power law with the same exponent independent of the value of the damping factor.

Obviously, equation (1) suggests that PageRank and In-Degree are intimately related, but this formula by itself does not explain the observed similarity in tail behavior. Furthermore, the linear algebra methods that have been commonly used in the PageRank literature [7,8] and proved very successful for designing efficient computational methods, seem to be insufficient for modelling and analyzing the asymptotic properties of the PageRank distribution.

The goal of our paper is to provide mathematical evidence for the power-law behavior of PageRank and its relation to the In-Degree distribution. Our approach is inspired by techniques from applied probability and stochastic operations research. The relation between PageRank and In-Degree is modelled through a distributional identity, which is analogous to the equation for the busy period in the M/G/1 queue (see e.g. [9]). Further, we analyze our model using the approach employed in [10] for studying the tail behavior of the busy period in case the service times are regularly varying random variables. This fits in our research because regular variation is in fact a formalization of the power law, and it has been widely used in queueing theory to model self-similarity, long-range dependence and heavy tails [11]. Thus, we use the notion of regular variation to model the power law distribution of In-Degree.

To obtain the tail behavior of PageRank in our model, we use Laplace-Stieltjes transforms and apply Tauberian theorems presented in the paper by Bingham and Doney [12], see also Theorem 8.1.6 in [13]. Even though our model contains some rather rigid simplifying assumptions – the most notable being independence between pages that link to the same page and a constant Out-Degree for all pages – these techniques allow us to prove the similarity in tail behavior for PageRank and In-Degree, thus suggesting that our assumptions do not touch upon the underlying reasons for this similarity. Moreover, our analysis allows to explicitly derive the multiplicative constant that quantifies the difference between PageRank and In-Degree tail behavior. Our analytical results show a good agreement with Web data.

2 Preliminaries

This section describes important properties of regularly varying random variables. We follow definitions and notations by Bingham and Doney [12], Meyer and Teugels [10], and Zwart [11]. More comprehensive details can be found in [13].

Definition 1. *A function L is said to be* slowly varying *if for every $t > 0$,*

$$\frac{L(tx)}{L(x)} \to 1 \quad as \quad x \to \infty.$$

Definition 2. *A random variable X is said to be* regularly varying *with index α if its distribution is such that*

$$\mathbb{P}(X > x) \sim x^{-\alpha} L(x) \quad as \quad x \to \infty,$$

for some positive slowly varying function $L(x)$. Here, as in the remainder of this paper, the notation $a(x) \sim b(x)$ means that $a(x)/b(x) \to 1$.

Denote by $f(s) = \mathbb{E}e^{-sX}$, $s > 0$, the Laplace-Stieltjes transform of X, and let $\xi_n = \mathbb{E}X^n$ be the nth moment of X, where $n \in \mathbb{N}$. The successive moments of X can be obtained by expanding f in a series at $s = 0$. More precisely, we have the following.

Lemma 1. *The nth moment of X is finite if and only if there exist numbers $\xi_0 = 1$ and $\xi_1, ..., \xi_n$, such that*

$$f(s) - \sum_{i=0}^{n} \frac{\xi_i}{i!}(-s)^i = o(s^n) \ as \ s \to 0.$$

If $\xi_n < \infty$ then we introduce the notation

$$f_n(s) = (-1)^{n+1} \left(f(s) - \sum_{i=0}^{n} \frac{\xi_i}{i!}(-s)^i \right). \tag{2}$$

Note 1. It follows from Lemma 1 that $\mathbb{E}X^n < \infty$ if and only if there exist numbers $\xi_0 = 1$ and $\xi_1, ..., \xi_n$ such that $f_n(s) = o(s^n)$ as $s \to 0$.

The following theorem establishes the relation between asymptotic behavior of a regularly varying distribution and its Laplace-Stieltjes transform. This result plays an essential role in our analysis.

Theorem 1. *(Tauberian Theorem) If $n \in \mathbb{N}$, $\xi_n < \infty$, $\alpha = n + \beta$, $\beta \in (0, 1)$, then the following are equivalent*

(i) $f_n(s) \sim (-1)^n \Gamma(1 - \alpha) s^\alpha L(\frac{1}{s})$ as $s \to 0$,
(ii) $\mathbb{P}(X > x) \sim x^{-\alpha} L(x)$ as $x \to \infty$.

Here and in the remainder of the paper we use the letter α to denote the index of the tail probability $\mathbb{P}(X > x)$.

3 Model

In this section we introduce a model that describes the relation between Page-Rank and In-Degree in the form of a stochastic equation. This model naturally follows from the definition of PageRank in (1), and is analytically tractable, thus enabling us to obtain the asymptotic behavior of PageRank. As will become clear, we make several rather strong simplifying assumptions. Nevertheless, the theoretical results of this model show a good match with observed Web graph behavior.

3.1 Relation between In-Degree and PageRank

Our goal now is to describe the relation between PageRank and In-Degree. To this end, we keep equation (1) almost unchanged but we transform it into a stochastic equation. Let R be the PageRank of a randomly chosen page. We treat R simply as a random variable whose distribution we want to determine. Further, we view the In-Degree of a random page as a random variable N, which follows a power law. The model for N will be specified in Section 3.2 below. In this work, we assume that the number of outgoing links (*Out-Degree*) $d \geq 1$ is the same for each page. This assumption is obviously not realistic; in particular it ignores the presence of 'hubs' (pages with extremely high Out-Degree) and 'dangling nodes' (pages with Out-Degree zero). The idea behind this rigid simplification is that we want to focus on the influence of the In-Degree, without considering other factors. Besides, it is a common belief that Out-Degrees do not affect the PageRank distribution, and it is also well-known (see e.g. [14]) that dangling nodes alter the PageRank vector only by a multiplicative constant. We note however that the proposed stochastic model allows for extensions. For instance, in the upcoming paper [15], we account for dangling nodes and allow for an arbitrary Out-Degree distribution.

Under the assumptions above, the random variable R satisfies a distributional identity

$$R \overset{d}{=} c \sum_{j=1}^{N} \frac{1}{d} R_j + (1 - c). \tag{3}$$

We now make the assumption that N and the R_j's are independent, and that the R_j's have the same distribution as R itself. We note that the independence assumption is not true in general. However, it is also not the case that the PageRank values of the pages linking to the same page i are directly related, so we may assume independence in this study.

The novelty of our approach is that we treat PageRank as a random variable which solves a certain stochastic equation. We believe, this approach is quite natural if our goal is to explain the power law behavior of PageRank because the power law is merely a description of a certain class of probability distributions. In fact, this point of view is in line with Pandurangan et al. [3] and other authors who consistently present log-log *histograms* of PageRank.

One of the nice features of the stochastic equation (3) is that it has the same form as the original formula (1). Thus, we may hope that our model correctly describes the relation between In-Degree and PageRank. This is easy to verify in the extreme (unrealistic) case when all pages have the same In-Degree d. In this situation, the PageRanks of all pages are equal, and it is easy to see that $R \equiv 1$ constitutes the unique solution of (3).

3.2   In-Degree Distribution

It is well-known that the In-Degree of Web pages follows a power law. For our analysis however we need a more formal description of this random variable, thus, we suggest to employ the theory of regular variation. We model the In-Degree of a randomly chosen page as a nonnegative, integer, regularly varying random variable, which is distributed as $N(X)$, where X is regularly varying with index α:

$$\mathbb{P}(X > x) \sim x^{-\alpha} L(x) \quad \text{as } x \to \infty,$$

and $N(x)$ is the number of Poisson arrivals on the time interval $[0, x]$. Without loss of generality, we assume that the rate of the Poisson process is equal to 1.

The advantage of this construction is that we do not need to impose any restrictions on X and at the same time ensure that the In-Degree is integer. It is intuitively clear that $N(X)$ is asymptotically equivalent to X, that is, $N(X)$ and X follow the same power law. Specifically, we have

$$\mathbb{P}(N(X) > x) \sim \mathbb{P}(X > x) \quad \text{as } x \to \infty. \tag{4}$$

For the proof of (4) using the Tauberian theorem (Theorem 1) see e.g. [1].

3.3   The Main Stochastic Equation

Combining the ideas from Sections 3.1 and 3.2, we arrive at the following equation

$$R \stackrel{d}{=} c \sum_{j=1}^{N(X)} \frac{1}{d} R_j + (1 - c), \tag{5}$$

where $c \in (0, 1)$ is the damping factor, $d \geq 1$ is the fixed Out-Degree of each page, and $N(X)$ describes the In-Degree of a randomly chosen page as the number of Poisson arrivals on a regularly varying time interval X. As we discussed above, stochastic equation (5) adequately captures several important aspects of the PageRank distribution and its relation to the In-Degree distribution. Moreover, our model is completely formalized, and thus we can apply analytical methods in order to derive the tail behavior of the random variable R representing PageRank.

Linear stochastic equations like (5) have a long history. In particular, (5) is similar to the famous equation that arises in the theory of branching processes and describes many real-life phenomena, for instance, the distribution of the busy period in the $M/G/1$ queue:

$$B \stackrel{d}{=} \sum_{i=1}^{N(S_1)} B_i + S_1,$$

where B is the distribution of the busy period (the time interval during which the queue is non-empty), S_1 is the service time of the customer that initiated the busy period, $N(S_1)$ is the number of Poisson arrivals during this service time and the B_i's are independent and distributed as B. We refer to [9] and other books on queueing theory for more details. Also, see Zwart [11] for an excellent detailed treatment of queues with regular variation, and specifically the busy period problem. We would like to add that our equation (5) is a special case in a rich class of stochastic recursive equations that were discussed in detail in the recent survey by Aldous and Bandyopadhyay [16].

This concludes the model description. The next step will be to use our model for providing a rigorous explanation of the indicated connection between the distributions of In-Degree and PageRank.

4 Analysis

The idea of our analysis is to write down an equation for the Laplace-Stieltjes transforms of X and R and then make use of the Tauberian theorem to prove that R is regularly varying with the same index as X. Since X and $N(X)$ are asymptotically equivalent, this will give us the desired similarity in tail behavior of the PageRank R and the In-Degree $N(X)$.

Let r be the the Laplace-Stieltjes transform of R. As a result of the assumptions from Section 3, we can use (5) to express r in terms of f, the Laplace-Stieltjes transform of X, as follows:

$$r(s) := \mathbb{E}e^{-sR} = e^{-s(1-c)}\mathbb{E}\left[\mathbb{E}\left[\exp\left(-s\frac{c}{d}\sum_{i=1}^{N(X)} R_i\right)\middle| N(X)\right]\right]$$

$$= e^{-s(1-c)}\mathbb{E}\left[\left(\mathbb{E}\left[\exp\left(-s\frac{c}{d}R_i\right)\right]\right)^{N(X)}\right]$$

$$= e^{-s(1-c)}\mathbb{E}\left[\mathbb{E}\left[\left(r\left(s\frac{c}{d}\right)\right)^{N(X)}\middle| X\right]\right]$$

$$= e^{-s(1-c)}\mathbb{E}\exp\left(-\left(1 - r\left(s\frac{c}{d}\right)\right)X\right) = e^{-s(1-c)}f\left(1 - r\left(\frac{c}{d}s\right)\right).$$

It can be shown that for the typical values $d > 1$ and $0 < c < 1$ the above equation has a unique solution $r(s)$ which is completely monotone and has $r(0) = 1$.

We start the analysis with providing the correspondence between existence of the n-th moments of X and R. We remind that $\xi_1, \ldots, \xi_n$ denote the first

n moments of X. Further, denote the first n moments of R by $\rho_1, \ldots, \rho_n$, and define

$$r_n(s) = (-1)^{n+1} \left(r(s) - \sum_{k=0}^{n} \frac{\rho_k}{k!} (-s)^k \right),$$

as in (2). Note that taking expectations on both sides of (5) we easily obtain $\mathbb{E}R = \rho_1 = 1$. This follows from the independence of $N(X)$ and the R_j's and the fact that $\mathbb{E}N(X) = \mathbb{E}X = \xi_1 = d$.

The next lemma holds.

Lemma 2. *The following are equivalent*

(i) $\xi_n < \infty$,
(ii) $\rho_n < \infty$.

Note 2. Similar as in Note 1, we can reformulate Lemma 2 as

$$f_n(s) = o(s^n) \quad \text{if and only if} \quad r_n(s) = o(s^n).$$

Note 3. Note that the stochastic inequality $R \overset{d}{>} (1 - c) \left(\frac{c}{d} N(X) + 1 \right)$ implies that the tail of the PageRank R is at least as heavy as the tail of the In-Degree $N(X)$.

The proof of Lemma 2 is quite lengthy and is therefore omitted. The interested reader is referred to the full version of this paper, see [1]. Same applies to the proof of Corollary 1 below.

Corollary 1. *The following holds:*

$$r_n(s) - dr_n \left(\frac{c}{d} s \right) = f_n(t) + O(t^{n+1}),$$

where $t = 1 - r \left(\frac{c}{d} s \right)$.

Now we are ready to explain the similarity between the In-Degree and PageRank distributions. Specifically, we show that the tail probabilities $\mathbb{P}(R > x)$ and $\mathbb{P}(N(X) > x)$ for PageRank and In-Degree, respectively, approximately differ by a multiplicative constant as x grows large. The next theorem formalizes this statement.

Theorem 2. *The following are equivalent*

(i) $\mathbb{P}(N(X) > x) \sim x^{-\alpha} L(x) \quad$ *as* $\quad x \to \infty$,
(ii) $\mathbb{P}(R > x) \sim \dfrac{c^\alpha}{d^\alpha - c^\alpha d} x^{-\alpha} L(x) \quad$ *as* $\quad x \to \infty$.

Proof.
$(i) \to (ii)$ From (i) and (4) it follows that

$$\mathbb{P}(X > x) \sim x^{-\alpha} L(x) \quad \text{as} \quad x \to \infty. \tag{6}$$

Theorem 1 also implies that (6) is equivalent to $f_n(t) \sim (-1)^n \Gamma(1-\alpha) t^\alpha L\left(\frac{1}{t}\right)$, where $t(s) = 1 - r\left(\frac{c}{d}s\right) \sim (c/d)s$ as $s \to 0$. Hence, by Corollary 1 we obtain

$$r_n(s) - d r_n\left(\frac{c}{d}s\right) \sim (-1)^n \Gamma(1-\alpha)\left(\frac{c}{d}\right)^\alpha s^\alpha L\left(\frac{1}{s}\right) \quad \text{as } s \to 0. \qquad (7)$$

Then also for every $k \geq 0$, as $s \to 0$, we have

$$r_n\left(\left(\frac{c}{d}\right)^k s\right) - d r_n\left(\left(\frac{c}{d}\right)^{k+1} s\right) \sim (-1)^n \Gamma(1-\alpha)\left(\frac{c}{d}\right)^\alpha \left(\frac{c}{d}\right)^{\alpha k} s^\alpha L\left(\frac{1}{\left(\frac{c}{d}\right)^k s}\right)$$

$$\sim (-1)^n \Gamma(1-\alpha)\left(\frac{c}{d}\right)^\alpha \left(\frac{c}{d}\right)^{\alpha k} s^\alpha L\left(\frac{1}{s}\right).$$

Next, we write $r_n(s)$ in the form of an infinite sum as follows:

$$r_n(s) = \sum_{k=0}^{\infty} d^k \left(r_n\left(\left(\frac{c}{d}\right)^k s\right) - d r_n\left(\left(\frac{c}{d}\right)^{k+1} s\right) \right).$$

From the above representation we obtain

$$r_n(s) \sim (-1)^n \Gamma(1-\alpha) \frac{d^\alpha}{d^\alpha - c^\alpha d} \left(\frac{c}{d}\right)^\alpha s^\alpha L\left(\frac{1}{s}\right) \quad \text{as } s \to 0.$$

Now we again invoke Theorem 1, which leads to (ii).
(ii) → (i) The proof follows easily from (ii) and Corollary 1.

Thus, we have shown that the asymptotic behaviors of PageRank and In-Degree differ by the multiplicative constant $\frac{c^\alpha}{d^\alpha - c^\alpha d}$, while the power law exponent remains the same. In the next section we will experimentally verify this result.

5 Numerical Results

We verified our findings by computing PageRank on the public data of the Stanford Web from [17]. To identify the power law behavior, we used cumulative log-log plots, which are much less noisy than histograms.

In order to compute the slope α, we used the following maximum likelihood estimator proposed by Newman [18]:

$$\alpha = 1 + n \left(\sum_{i=1}^{n} \ln \frac{x_i}{x_{min}} \right)^{-1}. \qquad (8)$$

Here the quantities x_i, $i = 1, \ldots, n$, are the measured values, and x_{min} usually corresponds to the smallest value of X for which the power law behavior is assumed to hold.

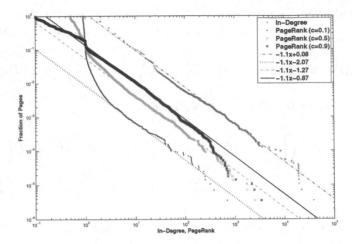

Fig. 1. Plots for the Web data. Fraction of pages with In-Degree/PageRank greater than x versus x in log-log scale, and the fitted straight lines.

There are several papers, see [3,4,5], and [6] that describe similar experiments for different domains and different number of pages, and they all confirm that PageRank and In-Degree follow power laws with the same exponent, around 1.1 for the cumulative distribution function.

We calculated all PageRank values for the Web graph with 281903 nodes (pages) and $\sim$ 2.3 million edges (links) using the standard power method (see e.g. [8]). On this dataset, the average Out-Degree, and hence average In-Degree is 8.2. In Figure 1 we show the log-log plots for In-Degree and PageRank of the Stanford Web Data, for different values of the damping factor ($c = 0.1$, 0.5 and 0.9). Clearly, these empirical values of In-Degree and PageRank constitute parallel straight lines for all values of the damping factor, provided that the PageRank values are reasonably large. It was observed in [6] that in general, PageRank depends on the damping factor but the PageRank of the top 10% of pages obeys a power law with the same exponent as the In-Degree, independent of the damping factor. This is in perfect agreement with our experimental results and the mathematical model, which is focused on the right tail behavior of the PageRank distribution.

The calculations based on the maximum likelihood method yield a slope -1.1, which verifies that In-Degree and PageRank have power laws with the same exponent $\alpha = 1.1$ (this corresponds to the well known value 2.1 for the histogram). More precisely, in Figure 1 we fitted the lines $y = -1.1x + 0.08$, $y = -1.1x - 0.87$, $y = -1.1x - 1.27$, and $y = -1.1x - 2.07$ to the plots of In-Degree and PageRank (with $c = 0.9$, $c = 0.5$ and $c = 0.1$, respectively).

We also investigated whether Theorem 2 correctly predicts the multiplicative constant

$$y(c) = \frac{c^\alpha}{d^\alpha - c^\alpha d}.$$

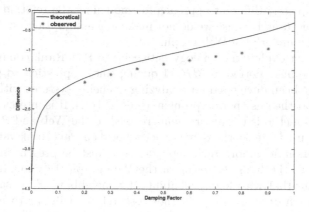

Fig. 2. The theoretical and observed differences between logarithmic asymptotics of In-Degree and PageRank

In Figure 2 we plotted $\log_{10}(y(c))$ and we compared it to the observed differences between the logarithms of the complementary cumulative distribution functions of PageRank and In-Degree, for different values of the damping factor. Obviously, in the data set, the assumption that all Out-Degrees are equal to some constant d is not satisfied. Therefore, we take $d = 8.2$, which is equal to the average In/Out-Degree in the Web data. As can be seen, the theoretical and observed values are quite close. E.g., for typical values of c between 0.8 and 0.9, the difference is 0.41, resulting in a factor $y(c)$ that is only a factor 2.57 larger than in the observed data. Thus, our model not only allows to prove the similarity in the power law behavior but also gives a good approximation for the difference between the two distributions.

The discrepancy between the predicted and observed values of the multiplicative constant suggests that our model does not capture PageRank behavior to the full extent. For instance, the assumption of the independence of PageRank values of pages that have a common neighbor may be too strong. We believe however that the achieved precision, especially for small values of c, is quite good for our relatively simple stochastic model.

6 Discussion

Our model and analysis resulted in the conclusion that PageRank and In-Degree should follow power laws with the same exponent. Growing Network models may provide an alternative explanation [19,20]. For instance, Avrachenkov and Lebedev [19] showed that in Growing Networks, introduced by Barabási and Albert [21], the *expected* PageRank follows a power law with an exponent, which does depend on the damping factor but equals ≈ 1.08 for $c = 0.85$. Note that our present model suggests that the power law exponent of PageRank does *not* depend on the damping factor. We emphasize that compared to [19,20], our model provides a completely different approach for modelling the relation between In-Degree and PageRank because we do not make any assumption on the structure or growth of the underlying Web graph.

We can further exploit the analogy between the PageRank equation and the equation for the busy period in $M/G/1$ queue, since sophisticated probabilistic techniques have been developed for analyzing queueing systems with heavy tails and in particular the busy period problem (see e.g. [11]). It is interesting to apply these advanced methods to the problems related to the Web and PageRank.

Our current model lacks the dependencies between PageRank values of pages sharing a common neighbor. Such dependencies must be present in the Web in particular due to the high clustering of the Web graph [18] (roughly speaking, clustering means that with high probability, two neighbors of the same page are connected to each other). In our further research we will try to include some sort of dependencies along with dangling nodes and random Out-Degrees [15]. Besides, we could also consider personalization or topic sensitivity [22]. The impact of these factors on the PageRank distribution could be determined by extending and generalizing the proposed analytical model.

Acknowledgment

Nelly Litvak gratefully acknowledges the financial support of the Netherlands Organization for Scientific Research (NWO) under the Meervoud grant 632.002.401.

References

1. Litvak, N., Scheinhardt, W.R.W., Volkovich, Y.: In-Degree and PageRank: Why do they follow similar power laws? (to appear in Internet Math.)
2. Brin, S., Page, L.: The anatomy of a large-scale hypertextual Web search engine. Computer Networks and ISDN Systems 33, 107–117 (1998)
3. Pandurangan, G., Raghavan, P., Upfal, E.: Using PageRank to characterize Web structure. In: H. Ibarra, O., Zhang, L. (eds.) COCOON 2002. LNCS, vol. 2387, Springer, Heidelberg (2002)
4. Donato, D., Laura, L., Leonardi, S., Millozi, S.: Large scale properties of the Webgraph. Eur. Phys. J. 38, 239–243 (2004)
5. Fortunato, S., Flammini, A., Menczer, F., Vespignani, A.: The egalitarian effect of search engines (2005), arxiv.org/cs/0511005
6. Becchetti, L., Castillo, C.: The distribution of PageRank follows a power-law only for particular values of the damping factor. In: Proceedings of the 15th international conference on World Wide Web, pp. 941–942. ACM Press, New York (2006)
7. Berkhin, P.: A survey on PageRank computing. Internet Math. 2, 73–120 (2005)
8. Langville, A.N., Meyer, C.D.: Deeper inside PageRank. Internet Math. 1, 335–380 (2003)
9. Robert, P.: Stochastic networks and queues. Springer, New York (2003)
10. Meyer, A.D., Teugels, J.L.: On the asymptotic behaviour of the distributions of the busy period and service time in M/G/1. J. App. Probab. 17, 802–813 (1980)
11. Zwart, A.P.: Queueing Systems with Heavy Tails. PhD thesis, Eindhoven University of Technology (2001)

12. Bingham, N.H., Doney, R.A.: Asymptotic properties of supercritical branching processes. I. The Galton-Watson process. Advances in Appl. Probability 6, 711–731 (1974)
13. Bingham, N.H., Goldie, C.M., Teugels, J.L.: Regular Variation. Cambridge University Press, Cambridge (1989)
14. Avrachenkov, K., Litvak, N., Nemirovsky, D., Osipova, N.: Monte Carlo methods in PageRank computation: When one iteration is sufficient (electronic). SIAM Journal on Numerical Analysis 45(2), 890–904 (2007)
15. Volkovich, Y., Litvak, N., Donato, D.: Determining factors behind the PageRank log-log plot. In: Bonato, A., Chung, F.R.K. (eds.) WAW 2007. LNCS, vol. 4863, Springer, Heidelberg (2007)
16. Aldous, D., Bandyopadhyay, A.: A survey of max-type recursive distributional equations. Ann. Appl. Probab. 15, 1047–1110 (2005)
17. Stanford dataset: (Accessed in March 2006), http://www.stanford.edu/simsdkamvar/research.html
18. Newman, M.E.J.: Power laws, Pareto distributions and Zipf's law. Contemporary Physics 46, 323–351 (2005)
19. Avrachenkov, K., Lebedev, D.: PageRank of scale free growing networks. Internet Mathematics 3(2), 207–231 (2006)
20. Fortunato, S., Flammini, A.: Random walks on directed networks: The case of PageRank (2006), arxiv.org/physics/0604203
21. Albert, R., Barabási, A.L.: Emergence of scaling in random networks. Science 286, 509–512 (1999)
22. Haveliwala, T.H.: Topic-sensitive PageRank. In: Proceedings of the Eleventh International World Wide Web Conference, Honolulu, Hawaii (2002)

Communities in Large Networks: Identification and Ranking

Martin Olsen

Department of Computer Science
University of Aarhus*
mo@daimi.au.dk

Abstract. We study the problem of identifying and ranking the members of a community in a very large network with link analysis only, given a set of representatives of the community. We define the concept of a *community* justified by a formal analysis of a simple model of the evolution of a directed graph. We show that the problem of deciding whether a non trivial community exists is NP complete. Nevertheless, experiments show that a very simple greedy approach can identify members of a community in the Danish part of the web graph with time complexity only dependent on the size of the found community and its immediate surroundings. The members are ranked with a "local" variant of the PageRank algorithm. Results are reported from successful experiments on identifying and ranking Danish Computer Science sites and Danish Chess pages using only a few representatives.

1   Introduction

A community in a network is a set of somewhat isolated vertices linking heavily to each other - for example a set of pages in the web graph related to a particular topic. People controlling a group of vertices (and their outgoing links) in a community are always looking for answers to the questions "How strong are the positions in the community for the members in my group?" and "How can these positions be improved?". The main objective for the work behind this paper is to establish a model of the community producing satisfactory answers to the first question. The model should also be small enough to enable a formal analysis leading to answers to the second question.

The purpose of the techniques in this paper is not to partition the network in to several communities. The purpose is to isolate and rank the members of a *single* community given by a set of representatives. Before the discussion of related work we would like to introduce the notation used in this paper.

In this paper $G = (V, E)$ denotes a directed graph where multiple occurrences of $(u, v) \in E$ are allowed. We will call $(u, v) \in E$ a *link* on u and say that u links to v etc. A link could for example represent a link from site u to site v in the

---

* The research is partly sponsored by the Danish company Cofman (www.cofman.com)

W. Aiello et al. (Eds.): WAW 2006, LNCS 4936, pp. 84–96, 2008.
© Springer-Verlag Berlin Heidelberg 2008

web graph or a reference in a paper written by u to a paper written by v. We define the *relative attention* that u shows v as $w_{uv} = \frac{m(u,v)}{outdeg(u)}$ where $m(u,v)$ is the multiplicity of link (u,v) in E. If $outdeg(u) = 0$ then $w_{uv} = 0$. For $C \subseteq V$ we let $w_{uC} = \sum_{c \in C} w_{uc}$, i.e. the attention that u shows the set of vertices C. In this paper we will reserve the term *edge* for an undirected graph.

1.1   Related Work

The problem of finding community structures in networks has been subject to a great deal of research - see e.g. [11].

Bagrow *et al.* [3] present a "local" method for detecting the community given by a single representative. A breadth first search from the representative stops when the number of edges connecting the visited vertices with un-visited vertices drops in a special way and reports the visited vertices as a community. Bagrow *et al.* repeat this procedure for each vertex and analyzes the overlap of the communities in order to eliminate problems with what the authors call "spill-over" of the breadth first search.

Formal definitions of communities are provided by Flake and different co-authors in [5] and [6]. According to [5] a community in an *undirected* graph with edges of unit capacity is a set of vertices C such that for all $v \in C$, v has at least as many edges connecting to vertices in C as it does to vertices in $\bar{C} = V - C$. Using the notion of relative attention extended to undirected graphs, this is $\forall v \in C : w_{vC} \geq \frac{1}{2}$. Flake *et al.* show in [5] how to identify a community containing a set of representatives as an *s-t* minimum cut in a graph with a virtual source s and virtual sink t. They show how the method can process only the neighborhood of the representatives yielding a local method with time complexity dependent on the size of the neighborhood. It is not possible for a vertex within a distance of more than two from the representatives to join the community for this "local" variant of their method.

The web graph is treated as a weighted *undirected* graph in [6] with an edge between page i and page j if and only if there is a link from page i to j or vice versa. Edge $\{i, j\}$ has weight $w_{ij} + w_{ji}$ following our definitions of attention. The graph is expanded with a virtual vertex t connected to all vertices with edges with the same weight α and the *community* of page s is defined by means of an *s-t* minimum cut. The members of such a community can be identified with a maximum flow algorithm.

The definitions in [5] and [6] are not based on a model of the evolution of a graph. It should also be noted that it seems impossible for a universally popular member to be a member of a small community by the definitions in [5] and [6]. A relatively high in-degree of a member will prevent it from being on the community side of a minimum cut. In fact any member v of a relatively small community in a relatively large network is risking being forced to leave the community if v attracts some attention from non community members if the community definition is based on minimum cuts and the graph is undirected.

Recently Andersen *et al.* [1] and Andersen and Lang [2] presented some very interesting approaches to identifying communities containing specific vertices. In both papers random walks are used to identify the communities. The graphs are assumed to be *unweighted* and *undirected* where this paper deals with *directed* graphs.

The search engine Google uses the PageRank algorithm [4,12] to calculate a universal measure of the popularity of the web pages. For a given search query the universal measure is combined with a measure of relevance with respect to the query in order to rank the web pages. Several variants of the PageRank algorithm have been proposed to make it personalized or topic/query specific - see for example [8,9,13].

1.2   Our Results

We present a community definition justified by a formal analysis of a very simple model of the evolution of a directed graph. We show that the problem of deciding whether a community $C \neq V$ exists such that $R \subseteq C$ for a given set of representatives R is NP complete. Nevertheless, we show that a fast and simple parameter free greedy approach performs well when detecting communities in the Danish part of the web graph. The time complexity of the approach is only dependent on the size of the found community and its immediate surroundings. Our method is "local" as the method in [3] but it does not use breadth first searches. We also show how to use a computationally inexpensive local variant of PageRank to rank the members of the communities and compare the ranking with the PageRank for the total graph.

These are two possible applications of the algorithms presented in this paper:

- Consider the following scenario: A user interested in Computer Science visits some sites on this subject. A piece of software running in the background finds that the Computer Science sites are similar by analyzing the content of the sites. It uses the Computer Science sites as the set R and reports a community C containing R with the sites ranked by our ranking algorithm. A real world example in Sect. 4.2 documents that this list could be very useful to the user!
- The ranking of the members of a community is the stationary probability distribution of a Markov Chain with the community as the state space. This Markov Chain can form the basis for an analysis leading to answers to questions like "Which link modifications would be optimal wrt. ranking for our group of nodes?".

In Sect. 2 the community definition and the greedy approach for identifying community members are presented. The ranking algorithm is introduced in Sect. 3 and the experiments are reported in Sect. 4.

2   Locating Communities

2.1   Community Definition

The intuition behind our community definition is that every community member shows more attention to the community than any non member:

Definition 1. *A community is a set $C \subseteq V$ such that*

$$\forall u \in C, \forall v \in \bar{C} : w_{uC} \geq w_{vC} \ .$$

Consider the following process: Assume the existence of a set $C \subset V$ and numbers p_1 and p_2 with $0 \leq p_1 < p_2 \leq 1$ such that the following holds: Every time a vertex $u \in C$ links to another vertex it will link to a member in C with probability p_2. Every time a vertex $v \in \bar{C}$ establishes a link it will link to a member in C with probability p_1. Each member of V establishes exactly q links independently of all other links established.

The number p_2 can be smaller than $\frac{1}{2}$ which means that the members of C does not necessarily predominantly link to other members of C as supposed in [5].

Definition 1 is justified by the following theorem:

Theorem 1. *Consider the process defined above and let $n = |V|$. If $\alpha = \left(1 - \frac{p_1}{p_2}\right) / \ln \frac{p_2}{p_1}$ then:*

$$P(\forall u \in C, \forall v \in \bar{C} : w_{uC} \geq w_{vC}) \geq 1 - n \left(\frac{e^{\alpha-1}}{\alpha^\alpha}\right)^{p_2 q} \ . \tag{1}$$

Proof. Let X_{xC} denote the number of links established by x linking to members in C. Let $\mu_2 = p_2 \cdot q$ denote the expected value for X_{uC} if $u \in C$. The expected value for X_{vC} for $v \in \bar{C}$ is $\mu_1 = p_1 \cdot q$.

We will establish an upper bound for the event in (1) *not happening*:

$$P(\exists u \in C, \exists v \in \bar{C} : X_{uC} < X_{vC}) \leq$$

$$P(\exists u \in C : X_{uC} < \tau \vee \exists v \in \bar{C} : X_{vC} > \tau) \leq$$

$$|C| \cdot P(X_{uC} < \tau) + |\bar{C}| \cdot P(X_{vC} > \tau) \ . \tag{2}$$

where u and v are generic elements in C and $\bar{C}$ respectively. This upper bound holds for any value of τ. The strategy of the proof is to find a τ such that the factors $P(X_{uC} < \tau)$ and $P(X_{vC} > \tau)$ have a low common upper bound.

We will use two Chernoff bounds and produce upper bounds on the factors in (2) assuming $\tau = \alpha \mu_2 = \frac{p_2}{p_1} \alpha \mu_1$ for $\alpha \in (\frac{p_1}{p_2}, 1)$:

$$P(X_{uC} < \alpha \mu_2) \leq e^{-\mu_2} \left(\frac{e^\alpha}{\alpha^\alpha}\right)^{\mu_2} \ . \tag{3}$$

$$P\left(X_{vC} > \frac{p_2}{p_1} \alpha \mu_1\right) \leq e^{-\mu_1} \left(\frac{e}{\frac{p_2}{p_1}\alpha}\right)^{\frac{p_2}{p_1}\alpha\mu_1} = e^{-\mu_1} \left(\frac{p_1}{p_2}\right)^{\alpha\mu_2} \left(\frac{e^\alpha}{\alpha^\alpha}\right)^{\mu_2} \ . \tag{4}$$

Now we will find a necessary and sufficient condition for these upper bounds to be identical:

$$e^{-\mu_2} = e^{-\mu_1} \left(\frac{p_1}{p_2}\right)^{\alpha\mu_2} \Leftrightarrow$$

$$-\mu_2 = -\mu_1 + \alpha\mu_2 \ln \frac{p_1}{p_2} \Leftrightarrow$$

$$\alpha = \left(1 - \frac{p_1}{p_2}\right) / \ln \frac{p_2}{p_1} .$$

The upper bounds in (3) and (4) are identical for this value of α which is easily shown to satisfy $\alpha \in (\frac{p_1}{p_2}, 1)$. We will put the common value $(\frac{e^{\alpha-1}}{\alpha^\alpha})^{\mu_2}$ in (2):

$$P(\exists u \in C, \exists v \in \bar{C} : X_{uC} < X_{vC}) \leq n \left(\frac{e^{\alpha-1}}{\alpha^\alpha}\right)^{p_2 q} .$$

□

Theorem 1 shows that real communities with $p_2 > p_1$ probably will obey Definition 1 in a large network where the number of links from each vertex is logarithmically lower bounded as pointed out by the following corollary:

Corollary 1. *For fixed p_1 and p_2 with $p_1 < p_2$ there exists a constant $k > 0$ such that*

$$P(\forall u \in C, \forall v \in \bar{C} : w_{uC} \geq w_{vC}) \to 1 \quad for \quad n \to \infty .$$

for $q = k \cdot \log n$.

Before addressing computability issues a couple of remarks on our community definition are in place. First of all there might be several communities containing a given set of representatives so picking the representatives might require several attempts. The experiments in Sect. 4.1 deal with the problem of choosing representatives. Secondly the union $C = C_1 \cup C_2$ of two communities C_1 and C_2 is not necessarily a community. For example there might be a vertex $v \in \bar{C}$ with $w_{vC} = 1$ and a vertex $u \in C$ with $w_{uC} < 1$ in which case C would not be a community since $w_{uC} < w_{vC}$. Communities in the "real world" seem to share these properties with our formal communities.

2.2 Intractability

We will now formally define the problem of deciding whether a non trivial community exists for a given set R:

Definition 2. *The COMMUNITY problem:*

- *Instance: A directed graph $G = (V, E)$ and a set of vertices $R \subset V$.*
- *Question: Does a community $C \neq V$ according to Definition 1 exist such that $R \subseteq C$?*

If we had an effective algorithm locating a non trivial community if at least one such community existed then we also could solve COMMUNITY effectively but even solving COMMUNITY effectively seems hard according to the following theorem:

Theorem 2. *COMMUNITY is NP complete.*

Proof. We can check in polynomial time whether C is a community containing R by calculating w_{xC} for all $x \in V$ thus COMMUNITY is in NP.

We will transform an instance of the NP complete problem PARTITION [7, page 223] into an equivalent instance of COMMUNITY in polynomial time. This means that we can solve the NP complete problem PARTITION in polynomial time if we can solve COMMUNITY in polynomial time thus COMMUNITY is NP complete since it is a member of NP. The rest of the proof contains the details of the transformation.

An instance of PARTITION is a finite set $A = \{a_1, a_2, \dots, a_n\}$ and a size $s(a_i) \in Z^+$ for each $a_i \in A$. The question is whether a subset $A' \subset A$ exists such that $\sum_{a \in A'} s(a) = \frac{S}{2}$ where S is the sum of the sizes of all elements in A? We will transform this instance into the instance of COMMUNITY given by a directed graph $G(V, E)$ with $n + 2$ vertices and $R = \{r\}$ where r is one of the vertices in G. The graph G is constructed in the following way:

We will start with two vertices r and y. For each $a_i \in A$ we will make a vertex with two links (a_i, r) and (a_i, y) with multiplicity 1 and two links (r, a_i) and (y, a_i) with multiplicity $s(a_i)$. The resulting network is shown on Fig. 1.

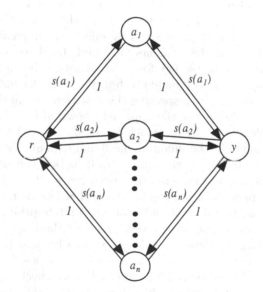

Fig. 1. A non trivial community C with $r \in C$ exists if and only it is possible to divide the set A in two parts with the same size. Each link is labeled with its multiplicity.

Now we will prove that G contains a non trivial community C containing R if and only if A' exists.

- If A' exists then $C = \{r\} \cup A'$ is a non trivial community containing r since $w_{xC} = \frac{1}{2}$ for all $x \in V$.

– Now assume that C is a non trivial community containing r. If C contains y then C also contains all the a's since $w_{aC} = 1$ if $\{r, y\} \subseteq C$. Since C is a non trivial community we have $y \notin C$. Now set $A' = C \cap A$.
  • If $\sum_{a \in A'} s(a) < \frac{S}{2}$ then $w_{rC} < \frac{1}{2}$ but there is at least one $a \notin C$ with $w_{aC} = \frac{1}{2}$ contradicting that C is a community.
  • If $\sum_{a \in A'} s(a) > \frac{S}{2}$ then $w_{yC} > \frac{1}{2}$ but there is at least one $a \in C$ with $w_{aC} = \frac{1}{2}$ - yet another contradiction.
We can conclude that $\sum_{a \in A'} s(a) = \frac{S}{2}$.   □

The network in Fig. 1 might be illustrative when comparing the definitions of a community in this paper and in [6]. If $A' \subset A$ exists such that $\sum_{a \in A'} s(a) = \sum_{a \in A-A'} s(a)$ then $C = \{r\} \cup A'$ will not be a community by the definition in [6] for any value of α.

2.3   A Greedy Approach

Despite the computational intractability experiments show that it is possible to find communities in the Danish part of the web graph with a simple greedy approach (see Sect. 4).

The approach starts with $C = R$. It then moves one element from $\bar{C}$ to C at a time choosing the element $v \in \bar{C}$ with the highest value of w_{vC}. After moving v to C it updates w_{xC} for all x linking to v and checks whether the current C satisfies Definition 1. The approach can be effectively implemented using two priority queues containing the elements in C and the elements in $\bar{C}$ linking to C respectively using w_{xC} as the key for x. The C-queue is a min-queue and the $\bar{C}$-queue is a max-queue. It is possible to find the next element to move and to decide if C is a community by inspecting the first elements in the queues as can be seen from the pseudo code of the approach shown in Fig. 2.

The time complexity of the approach is $O((n_C + m_C) \log n_C)$ where n_c is the number of elements in the union of the found community C and the set of vertices linking to C and m_C is the number of links between elements in C plus the number of links to C from $\bar{C}$ - multiple occurrences of $(u, v) \in E$ only counts as one link. The argument for the complexity is that less than n_C elements have to move between the two queues and that m_C update-priority operations are performed on the two queues containing no more than n_C elements. We are assuming that finding one vertex x linking to v can be done in constant time.

Some of the representatives might have no links, so we do not consider the attention shown by the representatives to C when we check whether C satisfies our definition of a community for the experiments in this paper. To be more specific we check whether

$$\forall u \in C - R, \forall v \in \bar{C} : w_{uC} \geq w_{vC} .$$

3   Ranking the Members

The PageRank algorithm used by Google can be viewed as a vote among *all* pages yielding a global measure of popularity. We will turn this into a vote among the

```
Greedy(G, R)
    C-queue := ∅
    C̄-queue := ∅
    forall r ∈ R do
        forall x ∈ V − R linking to r do
            if x ∈ C̄-queue then
                increase the priority of x with w_xr
            else
                insert x in the C̄-queue with priority w_xr
    while |C-queue| < minimum size or min(C-queue) < max(C̄-queue) do
        move the element v with maximum priority from the C̄-queue to the C-queue
        forall x ∈ V − R linking to v do
            if x ∈ C-queue or x ∈ C̄-queue then
                increase the priority of x with w_xv
            else
                insert x in the C̄-queue with priority w_xv
    Report R ∪ C-queue as a community
```

Fig. 2. Pseudo code for the greedy approach. Details for handling an empty C-queue or an empty $\bar{C}$-queue in the while-loop have been omitted for clarity.

relevant pages that are the pages in C. In this way we will obtain a small mathematical model which forms a basis for analyzing the consequences of changes in the link structure. The experiments carried out produces what we believe to be very valuable rankings which support the validity of the mathematical models behind the rankings.

A visitor to a community member $i \in C$ is assumed to have the following behavior:

– With probability given by some number d he decides to follow a link[1] from i. In this case there are two alternatives:
   • He decides to visit another member j of C. The probability that j gets a visit in this way is $d \cdot w_{ij}$.
   • He follows a link to a non member v. Assuming a low upper bound on w_{vC} it is not likely that the visitor will use a link to go back to C. Thus we treat this case as a jump to another member of C chosen uniformly at random.
– With probability $1 - d$ he decides to jump to another place without following a link which is treated as a jump to a member in C chosen uniformly at random.

A visitor to $i \in C$ will visit $j \in C$ with probability

$$p_{ij} = \frac{1-d}{|C|} + \frac{d(1-w_{iC})}{|C|} + d \cdot w_{ij} = \frac{1 - d \cdot w_{iC}}{|C|} + d \cdot w_{ij} .$$

---

[1] As suggested in [4] we use $d = 0.85$.

Like PageRank the ranking of the members is simply the unique stationary probability distribution of the Markov chain given by the transition matrix $P = \{p_{ij}\}_{i,j \in C}$. An iterative calculation of $\pi \cdot P^i$ will converge to the ranking in a few iterations where π is an arbitrary initial probability distribution. For details on convergence rates etc. we refer to the work of Langville and Meyer [10].

4 Experimental Work

For an on-line version of the results of the experiments please visit the home page of the author: www.daimi.au.dk/~mo/. Besides the results reported in this paper you can also find results from experiments with the s-t minimum cut approach from [6].

4.1 Identification of Community Members in Artificial Graphs

Inspired by Newman *et al.* [11] we test the greedy approach on some random computer generated graphs with known community structure. The graphs contain 128 vertices divided into four groups with 32 vertices each with vertices 1 - 32 in the first group, 33 - 64 in the next group etc. We will denote the first of the four groups as *group 1*. For each pair of vertices u and v either two links - (u, v) and (v, u) - or none are added to the graph. The pairs of links are placed independently at random such that the *expected* number of links from a vertex to vertices in the same group is 9 and the expected number of links to vertices outside the group is 7.

For 10 graphs the greedy approach reported the first community found containing at least 32 members with vertex number 1 as the single representative. The average size of the community found was 64.3 and the average number of vertices from group 1 in the community found was 28.9. If we use vertices 1 to 5 as representatives instead the corresponding numbers are 39.3 and 31.3 and if we use vertices 1 to 10 as representatives the numbers are 32.4 and 31.2. These admittedly few experiments suggest that the greedy approach can actually identify members of communities if the number of representatives is sufficient.

4.2 Identification and Ranking of Danish Computer Science Sites

Now we will demonstrate that the greedy approach is able to identify communities in the web graph using only a few representatives. A crawl of the Danish part of the web graph from April 2005 was used as the basis for the web experiments. In the first experiment reported in this paper V consists of the 180468 *sites* in the crawl where a link from site u to v is represented by $(u, v) \in E$.

The objective of the experiment was to identify and rank Danish Computer Science sites. The following four sites were used as representatives:

- **www.itu.dk**, IT University of Copenhagen
- **www.cs.auc.dk**, Department of Computer Science, University of Aalborg

Table 1. The Top 20 of two communities of Danish Computer Science sites. Representatives are written with bold font. The numbers after a site is the "global" ranking in the dk domain.

	556 members	1460 members
1	www.daimi.au.dk 267	www.au.dk 109
2	www.diku.dk 655	www.sdu.dk 108
3	**www.itu.dk** 918	www.daimi.au.dk 267
4	**www.cs.auc.dk** 1022	www.hum.au.dk 221
5	www.brics.dk 1132	www.diku.dk 655
6	**www.imm.dtu.dk** 1124	www.ifa.au.dk 681
7	www.dina.kvl.dk 1153	**www.itu.dk** 918
8	www.agrsci.dk 1219	www.ruc.dk 945
9	www.foejo.dk 1504	www.phys.au.dk 1051
10	www.darcof.dk 2113	www.brics.dk 1132
11	www.it-c.dk 2313	**www.cs.auc.dk** 1022
12	www.dina.dk 2169	www.dina.kvl.dk 1153
13	www.cs.aau.dk 2010	**www.imm.dtu.dk** 1124
14	rapwap.razor.dk 4585	www.agrsci.dk 1219
15	imv.au.dk 2121	www.kvinfo.dk 1122
16	razor.dk 2990	www.foejo.dk 1504
17	**www.imada.sdu.dk** 2998	www.bsd-dk.dk 1895
18	www.plbio.kvl.dk 3543	www.humaniora.sdu.dk 1826
19	www.math.ku.dk 2634	www.imv.au.dk 2121
20	mahjong.dk 3813	www.statsbiblioteket.dk 867

- **www.imm.dtu.dk**, Department of Informatics and Mathematical Modeling, Technical University of Denmark
- **www.imada.sdu.dk**, Department of Mathematics and Computer Science, University of Southern Denmark

The sites of the Departments of Computer Science for the two biggest universities in Denmark, **www.diku.dk** and **www.daimi.au.dk**, were *not included* in the set of representatives. These sites represent the universities in Copenhagen and Aarhus respectively.

The greedy approach found several communities. The Top 20 ranking of two communities with 556 and 1460 sites respectively are shown in Table 1 which also shows the ranking produced by a PageRank calculation on the dk domain. Members of both communities use more than 15-16 % of their links to other members and non members use less than 15-16 % on members.

The Top 20 lists contain mainly academic sites and the smaller community seems to be dominated by sites related to Computer Science. The ranking seems to reflect the "sizes" of the corresponding real world entities. It is worth noting that **www.daimi.au.dk** and **www.diku.dk** are ranked 1 and 2 in the smaller community. The site ranked 5 in the smaller community represents BRICS, Basic Research in Computer Science, which is an international PhD school within the areas of computer and information sciences, hosted by the Universities of Aarhus and Aalborg.

The larger community seems to be a more general academic community with the sites for University of Aarhus and University of Southern Denmark ranked 1 and 2 respectively. The larger community obviously contains the smaller community by the nature of the greedy approach.

The local ranking seems to reflect the global ranking with a few exceptions. The site rapwap.razor.dk is popular among the relevant sites but seems not to be that popular overall. The person behind rapwap.razor.dk has pages in Top 5 on Google searches[2] for Danish pages on "cygwin" and "php" which justifies rapwap.razor.dk's place on the Top 20 list of Danish Computer Science sites.

4.3   Identification and Ranking of Danish Chess Pages

We also carried out an experiment at the *page level* in order to rank Danish Chess pages using *one representative only*: www.dsu.dk, the homepage for the Danish Chess Federation. For this experiment V consisted of all pages up to three inter site links away from the representative where the links were considered unoriented. V contains approximately 330.000 pages. The weight w_{uv} is the fraction of inter site links on page u linking to page v.

The greedy approach located a community with 471 members. All members use at least 1.4 % of their inter site links on members and non members use less than 1.4 % on members. This means that only heavily linked non members link to the pages in the community and if they do they only link to the community with a few links. The Top 20 for this experiment – using the ranking from Sect. 3 – is shown in Table 2.

The page ranked 2 in the Top 20 is a page for a chess tournament held in Denmark in 2003 with several Grandmasters competing. The pages ranked 13 and 20 are pages (at that time) for the Danish and Scandinavian Chess championships respectively. Several of the subdivisions of the Danish Chess Federation (4, 7, 9, 19) are represented on the Top 20 and the page ranked 6 provides access to a database of more than 40.000 Chess games[3]. Most of the rest of the pages on the Top 20 are Chess Club pages. All in all the Top 20 seems useful from a Danish chess players point of view.

For comparison we searched Google[4] for Danish pages containing the word "skak" – the Danish word for chess. Several of the sites with pages in the Top 20 from Table 2 are also present in the Google search result but the latter seems targeted at a broader chess audience. The Google Top 20 contains for example several pages dealing with on-line chess and chess programs. The Top 20 from Table 2 seems to be targeted at a dedicated Danish chess player being a member of a chess club.

---

[2] The searches were carried out on January 23 2007.
[3] Appear to have moved to http://dsu9604.dsu.dk/partier/danbase.htm.
[4] The searches were carried out on April 12 2007.

Table 2. The top 20 of a community of 471 Danish chess pages found with the home-page of the Danish Chess Federation as a representative (written with bold font). The Danish word for chess is "skak".

1.	**www.dsu.dk**
2.	www.sis-mh-masters.dk
3.	dsus.dk
4.	www.8-hk.dk
5.	www.dsus.dk
6.	www.dsu.dk/partier/danbase.htm
7.	www.vikingskak.dk/4hk
8.	www.sk1968.dk
9.	www.4hk.dk
10.	www.skovlundeskakklub.dk
11.	www.vikingskak.dk
12.	www.alssundskak.dk
13.	www.skak-dm.dk
14.	www.frederikssundskakklub.dk
15.	www.birkeskak.dk
16.	home13.inet.tele.dk/dianalun
17.	www.rpiil.dk/nvf
18.	www.enpassant.dk/chess/index.html
19.	www.4hk.dk/index.htm
20.	www.skak-nm.dk

Acknowledgments. The author of this paper would like to thank Torsten Suel and his colleagues at Polytechnic University in New York for a crawl of the Danish part of the web graph and Gerth S. Brodal from University of Aarhus for valuable comments and constructive criticism.

References

1. Andersen, R., Chung, F.R.K., Lang, K.: Local graph partitioning using pagerank vectors. In: FOCS, pp. 475–486. IEEE Computer Society, Los Alamitos (2006)
2. Andersen, R., Lang, K.J.: Communities from seed sets. In: WWW 2006: Proceedings of the 15th international conference on World Wide Web, pp. 223–232. ACM Press, New York (2006)
3. Bagrow, J., Bollt, E.: A local method for detecting communities. Physical Review E 72, 046108 (2005)
4. Brin, S., Page, L.: The anatomy of a large-scale hypertextual Web search engine. Computer Networks and ISDN Systems 30(1–7), 107–117 (1998)
5. Flake, G., Lawrence, S., Giles, C.L.: Efficient identification of web communities. In: Sixth ACM SIGKDD International Conference on Knowledge Discovery and Data Mining, Boston, MA, August 20–23, pp. 150–160 (2000)
6. Flake, G., Tarjan, R., Tsioutsiouliklis, K.: Graph clustering and minimum cut trees. Internet Mathematics 1(4), 385–408 (2004)
7. Garey, M.R., Johnson, D.S.: Computers and Intractability: A Guide to the Theory of NP-Completeness. W. H. Freeman, New York (1979)

8. Haveliwala, T.H.: Topic-sensitive pagerank. In: WWW 2002: Proceedings of the 11th international conference on World Wide Web, pp. 517–526. ACM Press, New York (2002)
9. Jeh, G., Widom, J.: Scaling personalized web search. In: WWW 2003: Proceedings of the 12th international conference on World Wide Web, pp. 271–279. ACM Press, New York (2003)
10. Langville, A.N., Meyer, C.D.: Deeper inside pagerank. Internet Mathematics 1(3), 335–380 (2005)
11. Newman, M.E.J., Girvan, M.: Finding and evaluating community structure in networks. Physical Review E 69, 026113 (2004)
12. Page, L., Brin, S., Motwani, R., WinogradThe, T.: Pagerank citation ranking: Bringing order to the web. Technical report, Stanford Digital Library Technologies Project (1998)
13. Richardson, M., Domingos, P.: The Intelligent Surfer: Probabilistic Combination of Link and Content Information in PageRank. In: Advances in Neural Information Processing Systems 14, MIT Press, Cambridge (2002)

Combating Spamdexing: Incorporating Heuristics in Link-Based Ranking

Tony Abou-Assaleh and Tapajyoti Das

GenieKnows.com
Halifax, Nova Scotia, Canada
research@genieknows.com

Abstract. Users typically locate useful Web pages by querying a search engine. However, today's search engines are seriously threatened by malicious spam pages that attempt to subvert the unbiased searching and ranking services provided by the engines. Given the large fraction of Web traffic originating from search engine referrals and the high potential monetary value of this traffic, it is not surprising that some Web site owners try to influence the ranking function of a search engine in a malicious way, thus giving rise to Web spam. Since the algorithmic identification of spam is very difficult, most techniques require either some human assistance or extensive training to effectively deal with spam. We exploit the possibility of automatically reducing Web spam page in a Web collection by analyzing the Web graph, coupled with very simple content analysis. We present empirical evaluation of our approach on 1 million Web pages from the health domain. Our results clearly indicate that we can effectively filter out a significant fraction of Web spam pages.

1 Introduction

Search engines employ about 100 different features in the final ranking function of search results. These features can be grouped into two major categories: content-based features and link-based features. Many of the content-based features are used in the online ranking component to determine the relevancy of a Web page to a user query. Link-based features, on the other hand, are typically used in the offline ranking component to determine the overall quality of a page independent of a query.

Search engine optimization (SEO) is an IT industry that improves the structure, content, and presentation of Web sites to allow them to be better indexed and ranked by search engines. The intentions behind SEO techniques are legitimate: the goal is to facilitate locating relevant content to the end user.

Spamdexing[1] is an activity that attempts to artificially manipulate a page's ranking in a search engine [18]. The consequences of spamdexing with respect to a search engine can be detrimental. First, since there are financial advantages to be gained from search engine referrals, spam pages deprive legitimate Web

---

[1] The word spamdexing is a portmanteau of spamming and indexing.

W. Aiello et al. (Eds.): WAW 2006, LNCS 4936, pp. 97–106, 2008.

sites of the revenue. Secondly, they considerably deteriorate the quality of search engines by returning irrelevant results to the end user. As a result, search engines end up wasting a significant amount of their resources.

In this work, we exploit the possibility of automatically reducing spam documents by analyzing the Web graph, coupled with very simple content analysis. After presenting related work in Sec. 2 and reviewing essential related concepts in Sec. 3, we describe several simple and effective techniques for combating link-spam in Sec. 4. Section 5 details the evaluation and experimental setup and Sec. 6 presents and discusses the empirical results. We conclude the discussion in Sec. 7 and give our short and long-term future work directions.

2   Related Work

The earliest reference to using the term spamdexing is an article by Convey in 1996 [5]. At that time, most spamdexing was content based and tried to maliciously alter the content of a page to make it relevant for some queries. PageRank [14] was one of the first algorithm that used the Web graph to rank Web pages. PageRank tried to rank Web documents by assigning each document a global objective importance. Thus, it helped reduce content spam by emphasizing the importance of the link structure of the Web in ranking Web pages. However, it also opened the door to a new set of Web spamming techniques—link spam. Since there is no objective definition of spam, detecting link spammed pages is a difficult task. Moreover, pages participating in a link spam use more sophisticated techniques than content spamming making identifying these pages a nontrivial task.

Commercial search engines rarely reveal their anti-spam strategies. Literature on combating Web spam comes mostly from speculation by the SEO community [4,16] or from academic events. The International Workshop on Adversarial Information Retrieval on the Web was started in 2005 to bring together academics and practitioners interested in spamdexing.

There have been several attempts to classify Web spam, differing in the criteria and the level of details [9,18,4]. The popularity and effectiveness of these techniques rapidly changes as search engines devise new ways to combat spamdexing, while Web spammer devise new ways to exploit search engines.

A number of algorithms extend, modify, or complement PageRank to improve the ranking quality while reducing the rank of spam pages. For instance, Benczur [2] analyzed the distribution of PageRank scores of incoming links to identify spam page; Drost [6] applied machine learning to the spamdexing problem; Gyongyi [10] investigated the propagation of trust ranks from a manually selected good seed set, while Krishnan [11] looked at propagating anti-trust ranks from a spam seed set; and Wu [20] extended the trust rank idea to include topic information about the web pages. Other works focus on identifying features of a Web page that can differentiate spam pages from legitimate ones (a.k.a. ham pages) [13,17,19]. Another category of works analyzes link-spam strategies, their impact on PageRank, and their overall effectiveness [1,8].

3   Background

GenieKnows.com is developing a specialized search engine for the health domain[2]. Results are ranked by combining content and link-related features. In this work, we describe our efforts at reducing spamdexing from the top search results. We focus our discussion on combating link spam, although some techniques apply to content spamming as well.

3.1   PageRank

In this section we briefly discuss the PageRank [14] algorithm as the approach adopted by us primarily relies on PageRank. PageRank uses the link structure of the Web to assign a global importance score to all pages. The basic intuition behind PageRank is that a Web page is deemed important if other important Web pages point to it. Correspondingly, PageRank is based on a mutual reinforcement between pages; the importance of a certain page influences and is being influenced by the importance of other page.

The original summation-based definition is rarely used in the actual computation of PageRank. Instead, the following matrix definition is more common

$$PR = \alpha S + (1 - \alpha)E$$

where PR is the PageRank vector, $\alpha \in [0, 1]$ is a scaling factor that controls the relative importance of following a link vs. teleporting to a random page, S is a sparse, stochastic matrix that represents the link structure of the Web pages, and E is a teleportation vector (a.k.a. the personalization vector) [12].

3.2   Link Spamming

Link spam takes advantage of the link based ranking algorithms, such as PageRank, by artificially creating extraneous and often misleading links to boost the importance of one or more pages. Some common link spamming techniques are described below.

Link Farm. By creating tightly-knit communities (link farms) of Web pages referencing each other, the rank of each page in the farm is increased, and can be used to boost the ranking of a external target page.

Hidden Links. Strategically placing links where visitors will not be able to see them, thus increasing their hub score and boosting the ranks of the unrelated destination pages.

Honey Pots. In order to accumulate a number of incoming links to a single target page or a set of pages, spammers often create a set of pages that provide some useful resource (e.g., copies of Unix documentation pages), but that also have hidden links to some target pages. Most of the content in honey pots are usually copied of some other useful Web sites. The honey pot then attracts people to point to it, boosting indirectly the ranking of the target pages.

---

[2] Specialized search is often known by the name *vertical search* or *focused search*.

Buying Expired Domains. Some link spammers monitor DNS records for domains that will expire soon, then buy them when they expire and replace the pages with links to their own pages, thus taking the advantage of the false importance conveyed by the pool of old links.

Comment Spamming. Many Web sites have a field where anyone can post a comment (e.g. discussion boards, Wikis, and blogs); these sites usually do not ask for authentication to leave a comment. Spammers tend to post the URLs of their own Web sites thus getting an in-link from a good page. This kind of links defeat the purpose of PageRank and other link-based ranking algorithms, as this link is by no means a vote towards the spam site by the blog owner.

4   Combating Spamdexing

Since the algorithmic identification of spam is very difficult, most techniques significant manual effort or extensive training to efficiently deal with spamdexing. Our approaches minimizes the manual component by incorporating simple heuristics in link-based ranking to combat spamdexing.

We start with cleaning up the link graph. Next, we identify a core set of spamdexing pages. This set is further extended to include other likely spam pages. Finally, we use a biased PageRank-based ranking algorithm to produce the final off-line scores. The overall flow of our approach is depicted in Fig. 1. We discuss each of the stages below.

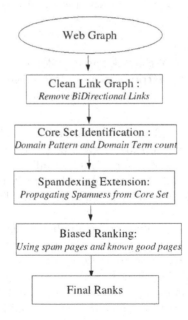

Fig. 1. The overall flow of our approach to combating spamdexing

4.1   Cleaning the Link Graph

We clean the entire link graph by removing all bi-directional links between do-main names (as opposed to Web pages). Additionally, we remove all internal links: links whose source and destination domain names or IP addresses are the same. Consequently, The presence of tightly-connected link farms and affiliated links are greatly reduced.

4.2   Core Set Identification

We use simple heuristics to identify a core set of Web spam pages. *Domain patterns* define common patterns of domain names that have high precision in identifying spam pages, such as *.biz, *.info, and *.pl. In [13] the authors re-port that 70% of all pages from *.biz domain to be spam. Certain domain names like *.biz, *.info, *.pl, *.us are easily available at a very cheap price, thus making them target for spammers. *Domain Term Count* marks as spam all pages whose domain names contain more than 5 terms; e.g., union-planters-bank. rbec-surf.sk The main motivation behind URL spamming is that many search engines pay special attention to words in host names and give these words a higher weight than if they occurred anywhere else in the Web page. Fetterly et. al. [7] observe that host names with many characters,dots and dashes are likely to be spam sites, which further validates our observation. Note that the domain patterns and the domain term count computations are highly scalable and inexpensive.

4.3   Spamdexing Extension

Next we extend the core Web-spam set using the *Spam Propagation* algorithm (Fig. 3). This algorithm is similar in its ideology to BadRank [16,19]. It starts with the core set, and at each iteration, it adds to the core set all pages that have links to a page in the core set. The process is repeated until the core set stabilizes. Hence, the spamness of a page is back-propagated to all pages that link to it, effectively capturing spam pages that are linked to the core set, but did not trigger the heuristic rules. Castillo et. al. [3] provide empirical evidence of the topological dependencies of spam pages. They show that non-spam pages tend to be linked by very few spam pages and usually link to non spam pages, while spam pages are mainly linked by spam pages. Figure 2 gives a graphical representation of how spamdexing extension works.

It is important to note that the Spam Propagation algorithm cannot be used to rank pages or sort them in any order. It is simply an attempt to detect pages which are possibly part of a link farm. As the algorithm iteratively detects spam pages by analyzing the in-links of a document, we might punish some legitimate pages that are victims of comment spam. One possible improvement for the Spam Propagation algorithm, would be to use a less strict penalty for the parent pages. We can use the *Parent Penalty* algorithm [19], where they use a threshold of 3 spam pages for propagation of the spam score.

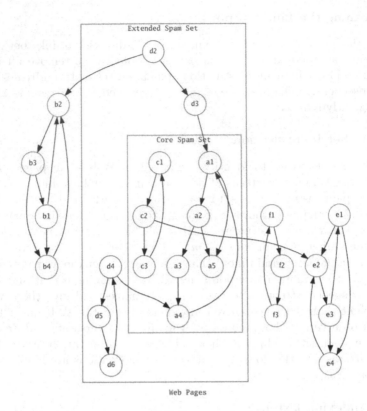

Fig. 2. Extending the core spam set. Nodes a_i and c_i are spam pages in the core set, d_i are pages identified as spam during spam propagation, and b_i, e_i, and f_i are pages that are treated as ham.

```
function SpamPropagation(C : core spam set )
begin
S := C
for each a ∈ S do
    for each b ∈ {p : p → a} do
        S := S ∪ {b}
return S
end
```

Fig. 3. Spam Propagation Algorithm

4.4   Biased Ranking

We remove all pages in the extended core set from the link graph, along with their in and out links. We manually select 100 non-spam pages from the vanilla PageRank ranking and favourably bias their initial rank proportionally to their PageRank ranking. We further uniformly bias all .gov and .edu pages under the assumption that a random page from these domains is less likely to point to a spam page than a random page from other domains.

Our biased ranking approach is similar to the TrustRank [10] algorithm in terms of selecting a seed set of good pages and using it to propagate the scores to the rest of the graph. However, in the TrustRank algorithm, uniform weights are assigned to all the documents in the seed set, and zero weights otherwise. In our biased rank, on the other hand, nonuniform weights are assigned to the seed set, proportional to their precomputed PageRank. We also assigned pages not classified under either of the two categories (extended spam set and good seed set) some initial weight, this was done to reduce the dominance of the .edu and .gov pages, and give advantage to the unclassified pages over the ones classified as spam. Moreover if we only propagate the weight from the seed set, some good quality pages that are not well connected to the seed set will be demoted. Pages classified as spam were assigned a zero weight. We compute the biased rank scores using 20 iterations and a scaling factor of 0.9.

5   Evaluation

We evaluate our methodology by a manual classification of 100 page samples from a Web page collection of 1 million pages from the health domain. The numbers reported represent precision only; calculating the recall measure would require a classification of the entire collection. Where appropriate, the sample is the top 100 ranked pages. Our testing corpus consists of 1 million pages collected from the health domain using our proprietary focused crawler.

We compute the accuracy of our three heuristics, domain patterns, domain term count, and spamdexing extension, in identifying spam pages. First, we apply each of the heuristic to the entire collection. Then, we selected a random sample of 100 pages. We manually inspect the pages in the sample and label them as spam or ham. Since there is no objective definition of spam, we followed the guidelines used in creating the WEBSPAM-UK2006 collection [15] for the evaluation. The percentage of the spam pages in the sample is the estimated accuracy of the heuristic.

We devise the link-based ranks of the Web pages in the collection using three ranking algorithms:

1. *Vanilla PageRank*—the PageRank algorithm is applied to the original, unprocessed collection;
2. *Clean PageRank*—the PageRank algorithm is applied to a "cleaned" collection where all the internal and bi-directional links are removed; and
3. *Biased Ranking*—the personalization vector in the PageRank algorithm is modified from being a uniform vector to a vector that is biased positively towards pages in the seed set (pages belonging to .edu and .gov domains) proportional to their Vanilla PageRank scores, negatively towards pages in the extended spam set, and neutral towards unclassified pages. The biased ranking algorithm is applied to the cleaned collection.

6   Experimental Results

The evaluation results are summarized in Table 1 and described below.

Table 1. Summary of Experimental Results

Algorithm	Spam	Ham	Accuracy	Comments
Vanilla PageRank	42%	58%	-	Top 100 pages
Domain Patterns	-	-	100%	Random sample of 100 pages
Domain Term Count	-	-	97%	Random sample of 100 pages
Spamdexing Extension	-	-	87%	Random sample of 100 pages
Clean PageRank	37%	63%	-	Top 100 pages
Biased Ranking	9%	91%	-	Top 100 pages

Vanilla PageRank, which does not include any of the methods discussed in section 4, contains about 42% of spam and 58% of ham pages. This is the baseline algorithm. The high percentage of spam pages in the top ranked documents renders the vanilla PageRank algorithm ineffective. In TrustRank [10], it is suggested to use the top ranked pages in page rank as the seed set for the TrustRank. Our finding show that this approach will not work with a collection like ours because a lot of high-ranked pages are spam.

Domain patterns heuristic have accuracy of 100% using a random sample, while the domain term count heuristic has accuracy of 97%, giving only 3% of false positives. The simplicity and high accuracy of these two heuristics makes them excellent candidates for identifying the core spam set. They are effective and efficient techniques that can be used to reduce spam in health Web pages.

Spamdexing extension has accuracy of 87%, giving 13% of false positives. Although the accuracy is not as high as in the previous two techniques, it is still sufficiently high to be useful. An examination of the false positives reveals that most of them are user-contributed online content (such as blogs, forums, and comments) that have been subjected to comment spam.

Clean PageRank is the vanilla PageRank applied to a the clean graph (i.e., bi-directional and internal links removed). The percentage of spam pages drops to 37%, compared to the baseline vanilla PageRank, reducing the spam by 5%. A 5% reduction in spam in the top 100 ranked documents is a significant improvement. Although some legitimate link may be removed during the cleaning process, the impact on spam pages is a lot more dramatic than on ham pages, which justifies removing the bi-directional and internal links.

Biased ranking is result of the final ranking stage. Only 9% of the top 100 pages are spam, giving a striking reduction of 33% over the baseline, and 28% over the clean PageRank. Note that although the algorithm is initially biased to favour pages from the .edu and .gov domains, only 10 pages from these domains appear in the top 100 pages.

7   Conclusion and Future Work

We present several simple and effective techniques to identify spam sites. We dramatically reduce the number of spam pages from the top ranked pages in PageRank without any significant increase in the computation complexity. We present an empirical evaluation of our approach on 1 million Web pages from the health domain. We were able to reduce the number of web spam pages in the top 100 ranked pages by 33% over the baseline PageRank.

The three heuristics presented, domain patterns, domain term count, and the spam propagation algorithm, are effective and efficient methods in reducing web spam. When combined with cleaning of the Web graph and a biased ranking algorithm, spam can be reduced by a significant factor.

We recognize that the presence of 9% spam in the top results after the biased ranking is still an alarming case. The methods presented in this paper act as first steps in combating spamdexing, but are insufficient by themselves in eliminating, or nearly eliminating spam pages from a collection.

Our short-term plans include evaluating this approach on a corpus of 20 million Web pages, incorporating other content-based heuristics to define the spamdexing core set, and revise the Biased Ranking algorithm to reduce the number of false positives.

Our long-term plans include combining other techniques for combating spamdexing such as detection of duplicate and near duplicate pages, link-farms, Topical PageRank, and the HITS hub-authority algorithm.

Acknowledgement

We thank Wennie Gao, Ashley George, and Yingbo Miao for their feedback and assistance in the evaluation. We thank Dr. Evangelos Milios for his valuable remarks.

References

1. Adali, S., Liu, T., Magdon-Ismail, M.: Optimal link bombs are uncoordinated. In: Proceedings of AIRWeb (2005)
2. Benczúr, A.A., Csalogány, K., Sarlós, T., Uher, M.: SpamRank – fully automatic link spam detection. Work in progress (2006)
3. Castillo, C., Donato, D., Gionis, A., Murdock, V., Silvestri, F.: Know your neighbors: Web spam detection using the web topology. Work in progress (2007)
4. Collins, G.: Latest Search Engine Spam Techniques. SitePoint (August 2004), http://www.sitepoint.com/print/search-engine-spam-techniques
5. Convey, E.: Porn sneaks way back on web. The Boston Herald, p. 028 (May 22, 1996)
6. Drost, I., Scheffer, T.: Thwarting the nigritude ultramarine: Learning to identify link spam. In: Proceedings of the European Conference on Machine Learning (2005)
7. Fetterly, D., Manasse, M., Najork, M.: Spam, damn spam, and statitics. In: Proceedings of WebDB 2004 (2004)

8. Gyöngyi, Z., Garcia-Molina, H.: Link spam alliances. In: Proceedings of the 31th VLDB Conference (2005)
9. Gyöngyi, Z., Garcia-Molina, H.: Web spam taxonomy. In: Proceedings of AIRWeb (2005)
10. Gyöngyi, Z., Garcia-Molina, H., Pedersen, J.: Combating web spam with TrustRank. In: Proceedings of the 30th VLDB Conference (2004)
11. Krishnan, V., Raj, R.: Web spam detection with anti-trust rank. In: Proceedings of AIRWeb (2006)
12. Langville, A.N., Meyer, C.D.: Google's PageRank and Beyond: The Science of Search Engine Rankings. Princeton University Press, Princeton (2006)
13. Ntoulas, A., Najork, M., Manasse, M., Fetterly, D.: Detecting spam web pages through content analysis. In: Proceedings of the WWW Conference (May 2006)
14. Page, L., Brin, S., Motwani, R., Winograd, T.: The pagerank citation ranking: Bringing order to the web. Technical report, Standford University (1999)
15. Yahoo! Research. Web Collection UK-2006. Yahoo! Research and University of Milan (2006), http://www.yr-bcn.es/webspam/datasets/uk2006-info/
16. Sobek, M.: PR0 - Google's PageRank 0 Penalty. eFactory GmbH & Co. KG Internet-Agentur (2003), http://pr.efactory.de/e-pr0.shtml
17. Urvoy, T., Lavergne, T., Filoche, P.: Tracking web spam with hidden style similarity. In: Proceedings of the AIRWeb (2006)
18. Wikipedia, the free encyclopedia. Spamdexing (August 2006), http://en.wikipedia.org/wiki/Spamdexing
19. Wu, B., Davison, B.D.: Cloaking and redirection: A preliminary study. In: Proceedings of AIRWeb (2005)
20. Wu, B., Goel, V., Davison, B.D.: Topical TrustRank: Using topicality to combat web spam. In: Procceddings of the WWW Conference (2006)

Traps and Pitfalls of Topic-Biased PageRank

Paolo Boldi[1,*], Roberto Posenato[2], Massimo Santini[1], and Sebastiano Vigna[1]

[1] Dipartimento di Scienze dell'Informazione, Università degli Studi di Milano, Italy
[2] Dipartimento di Informatica, Università degli Studi di Verona, Italy

Abstract. We discuss a number of issues in the definition, computation and comparison of PageRank values that have been addressed sparsely in the literature, often with contradictory approaches. We study the difference between *weakly* and *strongly* preferential PageRank, which patch the dangling nodes with different distributions, extending analytical formulae known for the strongly preferential case, and corroborating our results with experiments on a snapshot of 100 millions of pages of the .uk domain. The experiments show that the two PageRank versions are poorly correlated, and results about each one cannot be blindly applied to the other; moreover, our computations highlight some new concerns about the usage of exchange-based correlation indices (such as Kendall's τ) on approximated rankings.

1 Introduction

This paper started with an attempt to reproduce the correlation data published by Havelivala [1] about rankings biased towards different topics (where the correlation was computed using a measure similar to Kendall's τ); such seminal work has been receiving some attention lately, as in [2,3]. The bias was introduced using a *preference vector*, that is, by assuming that upon teleportation (see below for definitions) one does not land in a node chosen uniformly at random, but rather according to a given distribution.

During our attempts, we met significant difficulties due to the number of different ways in which PageRank can be defined and computed, and to the lack of public data over which to replicate the experiments. Following the incongruences in the literature, we were led to study in great detail the way in which PageRank depends on the preference vector and on the way dangling nodes are patched to obtain the final Markov chain. Also the way in which correlation indices are computed, and their depencence on the precision of the computation, turned out to be decisive.

We report the results obtained along our way. All our experiments use publicly available data gathered by UbiCrawler [4] on the .uk domain in the context of the EU project DELIS [5]. The topic-bias data we use are derived from the ODP [6] hierarchy. We believe such a public, well-defined data set is essential to continue research on personalised (and, in particular, topic-based) ranking.

First of all, we provide analytical formulae for *weakly preferential* and *strongly preferential* PageRank—two variants frequently found in the literature in which different distributions are used to patch dangling nodes. Using the Sherman–Morrison formula

---

* This work is partially supported by the EC Project DELIS and by MIUR PRIN Project "Automi e linguaggi formali: aspetti matematici e applicativi".

W. Aiello et al. (Eds.): WAW 2006, LNCS 4936, pp. 107–116, 2008.
© Springer-Verlag Berlin Heidelberg 2008

we are able to extend the results of Del Corso, Gullì and Romani [7] for strongly preferential PageRank. In doing so, we introduce the notion of *pseudorank*, a vector obtained using a PageRank-like matrix (which is not necessarily stochastic). Pseudoranks simplify greatly the following discussion, and present some interesting phenomena.

Then, we report experiments showing that weakly and strongly preferential PageRank can be very poorly correlated, and that results in the literature obtained using the two approaches are hardly comparable. In doing so, we use a low-level truncation technique that avoids the usual (and usually neglected in the literature) noise associated with the result of an interrupted iterative process, and we conclude by showing experimentally that such a noise may have a great impact on the computation of rank-based correlation indices.

2 PageRank

Albeit definitions of PageRank can be easily found in the literature, our purpose is precisely that of clarifying some relevant differences, so we start from scratch. Given a (web) graph G, the *row-normalised matrix* of G is the matrix P such that p_{ij} is one over the outdegree of i if there is an arc from i to j in G, zero otherwise. Note that in general P will not be stochastic, as it can have rows entirely made of zeroes.

Let us define d as the characteristic vector[1] of the dangling nodes (i.e., the vector with 1 in positions corresponding to nodes without outgoing arcs and 0 elsewhere). Let v and u be distributions[2], which we will call the *preference* and the *dangling-node* distribution.

PageRank r is defined (up to a scalar) by the eigenvector equation

$$r^T\big(\alpha(P + du^T) + (1 - \alpha)\mathbf{1}v^T\big) = r^T,$$

that is, as the stationary state of the Markov chain $\alpha(P + du^T) + (1 - \alpha)\mathbf{1}v^T$. More precisely, we have a *Markov chain with restart* [8] in which $P + du^T$ is the Markov chain (that follows the natural random walk on non-dangling nodes, and moves to a node at random with distribution u when starting from a dangling node) and v is the restart vector. The *damping factor* $\alpha \in [0 .. 1)$ decides how often the Markov chain follows the graph, and how often it moves at a random node following the preference vector v. The latter behaviour is commonly called *teleportation*, referring to a well-known random-walk metaphore in which a random surfer with probability α moves along an outlink chosen uniformly at random (or, in case of a dangling node, chosen among all nodes according to distribution u), and teleports to a random node chosen with distribution v with probability $1 - \alpha$. In the random-surfer metaphore, PageRank is the average fraction of time the surfer spends at a given node.

3 Strongly vs. Weakly Preferential

A significant amount of recent research is devoted to studying the dependence of PageRank on the preference vector. The preference vector biases the rank towards nodes

---

[1] All vectors in this work are column vectors.

[2] By *distribution* we mean a vector with non-negative entries and ℓ_1-norm equal to 1.

that are closer to nodes with a larger value in the preference vector. The preference vector, for instance, might depend on the user's preferences, in which case one speaks of *personalised PageRank* [2].

Clearly, the preference vector v significantly conditions PageRank. Some care must be exercised, however: real-world snapshots comprise a significant percentage of *dangling nodes* (nodes without outlinks), in particular if the graph contains the whole *frontier* of the crawl [9], rather than just the visited nodes. Hence, the way in which the surfer chooses the next node when she is at a dangling node (i.e., the choice of u) is also very relevant, and it is an issue resolved in different ways by different authors.

We distinguish clearly between *strongly preferential* PageRank, in which the preference and dangling-node distributions are identical (i.e., $u = v$), and correspond to a topic or personalisation bias, and *weakly preferential* PageRank, in which the preference and the dangling-node distributions are not identical, and, in principle, uncorrelated (most commonly, $u = 1/n$). As we shall see, the distinction is not irrelevant, as the correlation between weakly and strongly preferential PageRank can be quite low.

As a first analytical step to understand fully the relationship between preference and dangling-node distributions we extend the closed formula given by Del Corso, Gullì and Romani [7] for strongly preferential PageRank to a general formula that applies also to weakly preferential PageRank. Using this formula, any biased, weakly preferential PageRank vector whose distributions are a linear combination of a set of base vectors [2] can be computed using the *pseudorank* vectors associated to the base vectors. The computation of a pseudorank vector requires the same amount of computational effort as for computing PageRank, but once pseudoranks have been computed it is immediate to compute and compare several different biased ranks.[3]

PageRank r is defined (up to a scalar) by the eigenvector equation

$$r^T\left(\alpha(P + du^T) + (1 - \alpha)1v^T\right) = r^T.$$

After a transposition, imposing $r^T1 = 1$ and solving for r, we obtain the standard closed form

$$r = (1 - \alpha)\left(I - \alpha P^T - \alpha ud^T\right)^{-1}v.$$

The interesting point of this form is that it exhibits PageRank as a *linear operator* on the preference vector v.

Definition 1. *Let P be a row-normalised web-graph matrix. The* pseudorank *of P with preference vector v and damping factor $\alpha \in [0 .. 1]$ is defined as*

$$\widetilde{v}(\alpha) = (1 - \alpha)\left(I - \alpha P^T\right)^{-1}v.$$

We note by passing that if $d = 0$ (equivalently, if P is stochastic) then $\widetilde{v}(\alpha)$ is actually the PageRank.[4]

The above definition can be extended by continuity to $\alpha = 1$, albeit the fact is not trivial. The *resolvent* of a matrix M is the linear operator $\mathscr{R}(\mu, M) = (\mu I - M)^{-1}$,

---

[3] Actually, some papers, such as [3], use tacitly pseudoranks as the *definition* of PageRank.
[4] The condition number in the computation of pseudoranks is the same as or better than that for PageRank, and there is no increase of computation time.

defined for every μ which is not an eigenvalue of M; it can be expanded into a *Laurent series* around every eigenvalue of M [10,11]. In particular, the expansion around 1 is

$$\mathscr{R}(\mu, M) = \frac{M^*}{\mu - 1} + \sum_{k=0}^{\infty} (\mu - 1)^k Q^{k+1}$$

for a suitable matrix Q, where M^* is the *Cesáro limit*

$$M^* = \lim_{n \to \infty} \frac{1}{n} \sum_{k=0}^{n-1} M^k,$$

which is always defined and is equal to $\lim_{k \to \infty} M^k$ whenever the latter is defined [12]. This implies that

$$\lim_{\mu \to 1^+} (1 - \mu)\mathscr{R}(\mu, M) = M^*,$$

so the pseudorank for $\alpha \to 1$ is simply $(P^*)^T v$.

Theorem 1. *Let P the row-normalised matrix of a web graph, v the preference vector, u the dangling distribution and α the damping factor. Then, the PageRank vector r satisfies*

$$r = \tilde{v}(\alpha) - \tilde{u}(\alpha)\frac{d^T \tilde{v}(\alpha)}{1 - \frac{1}{\alpha} + d^T \tilde{u}(\alpha)}.$$

Proof. Let $\tilde{u}(\alpha)$ and $\tilde{v}(\alpha)$ be the pseudoranks of u and v, and define $R = I - \alpha P^T$. By the Sherman–Morrison formula [7], we have

$$r = (1 - \alpha)(R - \alpha u d^T)^{-1} v = (1 - \alpha)R^{-1}v + (1 - \alpha)\frac{R^{-1}u d^T R^{-1}}{\frac{1}{\alpha} - d^T R^{-1}u}v =$$

$$= \tilde{v}(\alpha) + \frac{\tilde{u}(\alpha)d^T \tilde{v}(\alpha)}{\frac{1}{\alpha} - 1 - d^T \tilde{u}(\alpha)}. \qquad \blacksquare$$

Note that the scalar values $d^T \tilde{v}(\alpha)$ and $d^T \tilde{u}(\alpha)$ have two very simple interpretation— they are the pseudorank accumulated by dangling nodes w.r.t. v and u, respectively.

By properly ordering multiplications, no matrix computation is necessary to compute the formula above. When $u = v$, the formula reduces to the one provided in [7]:[5]

$$r = \tilde{v}(\alpha) \left(1 - \frac{d^T \tilde{v}(\alpha)}{1 - \frac{1}{\alpha} + d^T \tilde{v}(\alpha)} \right) \qquad (1)$$

and the (rather surprising) consequence is that *pseudoranks are just multiples of strongly preferential ranks*. In other words, PageRank might as well be computed *without taking care of dangling nodes* by using the standard expansion

$$\tilde{v}(\alpha) = (1 - \alpha)(I - \alpha P^T)^{-1}v = (1 - \alpha)\sum_{n=0}^{\infty} \alpha^n (P^T)^n v.$$

---

[5] The reader should note that our formula has some difference in signs w.r.t. the original paper, where it was calculated incorrectly.

Indeed, by truncating the infinite sum we obtain an approximation of the pseudorank:

$$\left\| \widetilde{v}(\alpha) - (1 - \alpha) \sum_{n=0}^{k} \alpha^n \left(P^T\right)^n v \right\| = \left\| (1 - \alpha) \sum_{n=k+1}^{\infty} \alpha^n \left(P^T\right)^n v \right\| \leq \alpha^{k+1}.$$

The fact that the above formula approximates well PageRank up to a constant factor shows that actually PageRank is related more to a *diffusion* than to a *mixing* phenomenon. In other words, even if the PageRank definition is in term of Markov chains, its value can be computed also by a cumulative process in which the preference vector is broadcast to the neighbours using a decay factor α.

Pseudoranks are computed from their preference vector using a linear operator: as a consequence, both weakly and strongly preferential PageRank are quickly computable if, for instance, $\widetilde{e}_i(\alpha)$ is known for some base e_i of the vector space. This property is noted in [2] for strongly preferential PageRank, but Theorem 1 shows that the statement is true also in the weakly preferential case, albeit the dependence on u *is not linear*, so weakly preferential PageRank vectors do not obey the simple linear laws for what matters the dangling node distribution.

3.1 A Worked Example, and Some Observations

Let us consider the simple example of a graph with two nodes, and a single arc going from the first to the second node. With an arbitrary norm-one vector $x = (x, 1 - x)^T$ we have

$$\widetilde{x}(\alpha) = (1 - \alpha)(I - \alpha P^T)^{-1} x = \begin{pmatrix} (\alpha - 1)x \\ (1 - \alpha)(1 + (\alpha - 1)x) \end{pmatrix},$$

and $d^T \widetilde{x}(\alpha) = (\alpha - 1)x$. Note that, for every preference vector v, $\lim_{\alpha \to 1^-} d^T \widetilde{v}(\alpha) = 0$, and this is not by chance: $d^T \widetilde{v}(\alpha)$ has a limit as $\alpha \to 1$ because it is a rational function of α, and looking at (1) it is clear that $d^T \widetilde{v}(\alpha)$ cannot converge to any limit different from 0, or otherwise the strongly preferential PageRank would itself converge to the zero vector.

To complete the example, for two arbitrary norm-one vectors $v = (v, 1 - v)^T$ and $u = (u, 1 - u)^T$ we have that the denominator in Theorem 1 evaluates to $(\alpha - 1)(1/\alpha + u)$, giving

$$r = \frac{1}{\alpha u + 1} \begin{pmatrix} v + \alpha(u - v) \\ (\alpha - 1)v + 1 \end{pmatrix}.$$

The limit when α approaches 1 is

$$\lim_{\alpha \to 1^-} r = \frac{1}{1 + u} \begin{pmatrix} 1 \\ u \end{pmatrix}.$$

A precise estimate of the difference between strongly and weakly preferential PageRank can be obtained by considering the difference $r_1 - r_2$ between the rank values of the two nodes:

$$r_1 - r_2 = \frac{2(1 - \alpha)v + \alpha u - 1}{\alpha u + 1};$$

a contour plot of this difference for the strongly preferential and weakly preferential cases (as a function of α and v) is given in Figure 1: note that the behaviours in the two cases are significantly different, in particular when $\alpha > 0.5$ (an area that is quite important, since $\alpha = .85$ is the value that is customarily adopted for PageRank computation).

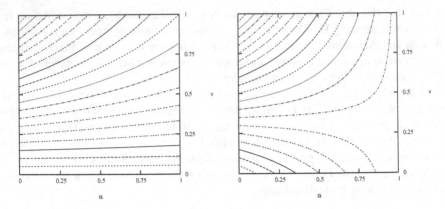

Fig. 1. The contour plot of $r_1 - r_2$ for the strongly preferential (left) and weakly preferential (right) case for the worked example

4 Experiments

Is the difference between strongly and weakly preferential significant also when only ranks are considered instead of rank values? To answer this question, we ran a number of experiments on a crawl of about 100 million pages of the .uk domain gathered for the DELIS project [5]. For comparisons we used Kendall's τ, a classical nonparametric correlation index that has recently received much attention within the web community for its possible applications to rank aggregation [13,14,15,16] and for determining the convergence speed in the computation of PageRank [17]. Here we follow exactly the definition given in [16].

In Figure 2 we show the values of Kendall's τ (in dependence of α) for weakly and strongly preferential PageRank where the preference distribution is in one case concentrated on a single node, and in the other case it is uniformly distributed among the nodes in the Open Directory Project [6] "Business" category. In both cases, u was set to the uniform distribution. The correlation between the two values is always very low (except, of course, when $\alpha \approx 0$).

The interesting phenomenon is that correlation *increases* as α increases, whereas intuition would suggest the opposite behaviour: as α increases, the graph becomes more relevant, so the structural differences between using v or the uniform distribution to patch dangling nodes should be more visible.

To understand whether the low correlation is due to topic concentration or to the number of nonzero entries, in Figure 3 we show the comparison of the same kind of values calculated on UK-2006 graph using an additional vector obtained by randomly

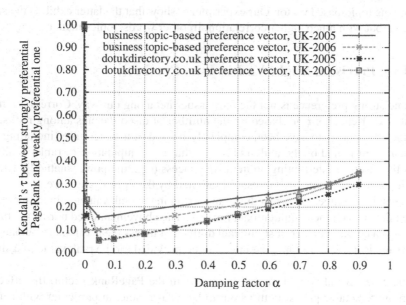

Fig. 2. Kendall's τ between weakly and strongly preferential PageRank computed on the UK-2005 and UK-2006 graphs using the ODP "business" topic-based preference vector and the `http://dotuk.directory.co.uk/` page-based preference vector

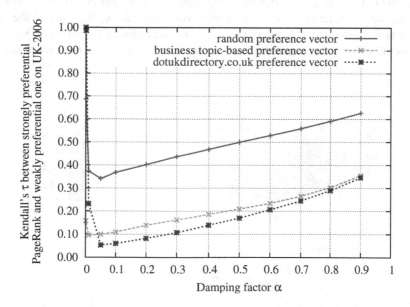

Fig. 3. Kendall's τ between weakly and strongly preferential PageRanks calculated on UK-2006 using three different preference vectors

shuffling the topic-based vector. Our experiments show that the latter exhibits the same behaviour, but with much higher correlation. In other words, topics matter.

5   More Precision Might End in Less Precision

Weak and strong preference is not the only issue met along our way. Correlation measures such as Kendall's τ are based on the number of discordancies among ranks, but the point that appears to have been completely missed in the literature (including that previously contributed by the authors [16]) is that the computation of ranks is almost always the result of interrupting an iterative process (e.g., the power method). The interruption is usually based on a threshold satisfied by the ℓ_1 or ℓ_2 measure.

As a result, a number of correct digits appearing in the rank values is hard to predict, as it just depends on the computational process. The abovementioned norms guarantee on *average* a certain number of significant digits, but unless the much more demanding ℓ_∞ measure is used, almost no guarantee can be provided for the rank value of a single node.

In the case several very close values appear in the PageRank vector, the effect of such an unpredictable precision turns out to be catastrophic, in particular with certain computational methods (such as Gauss-Seidel). Namely, the value of Kendall's τ is strongly influenced by the number of significant digits considered in its computation.

To prove the impact of this observation experimentally, we present data obtained by working out the strongly preferential PageRank computation in a standard fashion, using the Gauss-Seidel method, for a certain preference vector. We stopped the computation at different stages, having every time a known (lower bound on the) number of correct digits in the computed ranks, that we denote by p, and then we computed

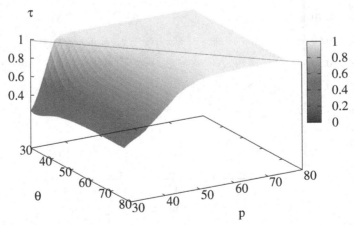

Fig. 4. Values of Kendall's τ: when rank values are batched using more bits than the number of significant bits guaranteed by the PageRank computation, the value of τ drops significantly. These data are determineted from the .uk web graph, using the ODP "adult" topic-based preference vector, $\alpha = .85$ and the Gauss-Seidel method. p is the number of correct binary digits, θ is the number of digits used to determine τ.

Kendall's τ using only a limited number of digits in the ranks. To limit the number of significant digits we used, we turned each floating point-number into its bitwise IEEE 754 representation, and manipulated it directly so to delete all digits beyond a certain threshold. This procedure, applied with threshold θ, has the effect of *batching* all values in the interval $[\, j2^{-\theta} \, .. \, (j+1)2^{-\theta}\,)$, into the value $j2^{-\theta}$. The net effect is that several rank values that appeared to be discordant because of unpredictable noise in the last digits are now considered as concordant. (We remark that due to the size of the data we use, these computations require thousands of hours of CPU time.)

The resulting graphs (an example is presented in Figure 4) are quite surprising: even the τ of a certain PageRank *computed against the same vector, but with a different precision* can go down as low as 0.2. Of course, as far as the computation of τ uses no more digits than those that are guaranteed to be correct, the correlation is 1, but it rapidly drops as soon as more digits are considered; in particular, computing τ blindly (i.e., without any form of batching) can bring essentially to random results. In a slogan: *more precision might end in less precision*. One must be always careful about the actual number of significant digits of each rank—using an ℓ_∞-measure guaranteeing the number of digits used in the computation of correlation indices is a safe choice.

More evidence is needed to corroborate the data we present. But already our preliminary results show that order-based correlation indices must be managed with great care, and have probably given rise to biased results in the past.

References

1. Haveliwala, T.H.: Topic-sensitive PageRank. In: The eleventh International Conference on World Wide Web Conference, pp. 517–526. ACM Press, New York (2002)
2. Jeh, G., Widom, J.: Scaling personalized web search. In: WWW 2003: Proceedings of the 12th international conference on World Wide Web, pp. 271–279. ACM Press, New York (2003)
3. Csalogány, K., Fogaras, D., Rácz, B., Sarlós, T.: Towards scaling fully personalized PageRank: Algorithms, lower bounds, and experiments. Internet Math. 2, 333–358 (2006)
4. Boldi, P., Codenotti, B., Santini, M., Vigna, S.: Ubicrawler: A scalable fully distributed web crawler. Software: Practice & Experience 34, 711–726 (2004)
5. DELIS: Dynamically Evolving Large-scale Information Systems EC FP6 project, http://delis.upb.de/
6. ODP: Open Directory Project, http://dmoz.org/
7. Del Corso, G., Gullì, A., Romani, F.: Fast PageRank computation via a sparse linear system. Internet Math. 2 (2006)
8. Boldi, P., Lonati, V., Santini, M., Vigna, S.: Graph fibrations, graph isomorphism, and PageRank. RAIRO Inform. Théor 40, 227–253 (2006)
9. Eiron, N., McCurley, K.S., Tomlin, J.A.: Ranking the web frontier. In: Proceedings of the 13th conference on World Wide Web, pp. 309–318. ACM Press, New York (2004)
10. Lasserre, J.B.: A formula for singular perturbations of Markov chains. Journal of Applied Probability 31, 829–833 (1994)
11. Yosida, K.: Functional Analysis, 6th edn. Springer, Heidelberg (1980)
12. Iosifescu, M.: Finite Markov Processes and Their Applications. John Wiley & Sons, Chichester (1980)

13. Fagin, R., Kumar, R., Sivakumar, D.: Comparing top k lists. In: Proceedings of the fourteenth annual ACM-SIAM symposium on Discrete algorithms, Society for Industrial and Applied Mathematics, pp. 28–36 (2003)
14. Fagin, R., Kumar, R., McCurley, K.S., Novak, J., Sivakumar, D., Tomlin, J.A., Williamson, D.P.: Searching the workplace web. In: Proceedings of the twelfth international conference on World Wide Web, pp. 366–375. ACM Press, New York (2003)
15. Dwork, C., Kumar, R., Naor, M., Sivakumar, D.: Rank aggregation methods for the web. In: Proceedings of the tenth international conference on World Wide Web, pp. 613–622. ACM Press, New York (2001)
16. Boldi, P., Santini, M., Vigna, S.: Do your worst to make the best: Paradoxical effects in PageRank incremental computations. Internet Math. 2, 387–404 (2005)
17. Kamvar, S.D., Haveliwala, T.H., Manning, C.D., Golub, G.H.: Extrapolation methods for accelerating pagerank computations. In: Proceedings of the twelfth international conference on World Wide Web, pp. 261–270. ACM Press, New York (2003)

A Scalable Multilevel Algorithm for Graph Clustering and Community Structure Detection

Hristo N. Djidjev[1,*]

Los Alamos National Laboratory, Los Alamos, NM 87545

Abstract. One of the most useful measures of cluster quality is the modularity of the partition, which measures the difference between the number of the edges joining vertices from the same cluster and the expected number of such edges in a random (unstructured) graph. In this paper we show that the problem of finding a partition maximizing the modularity of a given graph G can be reduced to a minimum weighted cut problem on a complete graph with the same vertices as G. We then show that the resulted minimum cut problem can be efficiently solved with existing software for graph partitioning and that our algorithm finds clusterings of a better quality and much faster than the existing clustering algorithms.

1 Introduction

One way to analyze and understand the information contained in the huge amount of data available on the WWW and the relationships between the individual items is to organize them into "communities," maximal groups of related items. Determining the communities is of great theoretical and practical importance since they correspond to entities such as collaboration networks, online social networks, scientific publications or news stories on a given topic, related commercial items, etc. Communities also arise in other types of networks such as computer and communication networks (the Internet, ad-hoc networks) and biological networks (protein interaction networks, genetic networks).

The problem of identifying communities in a network is usually modeled as a *graph clustering* (GC) problem, where vertices correspond to individual items and edges describe relationships. Then the communities correspond to clusters with a lot of edges between vertices belonging to the same subgraph (called *in-cluster* edges) and fewer edges between vertices from different subgraphs (called *between-cluster* edges). The GC problem has been intensively studied in the recent years in relation to its applications in the analysis of networks. Girvan and Newman propose in [11], [18] algorithms based on the *betweenness* of the edges of a graph, a characteristic that measures the number of the shortest paths in a graph that use any given edge. In [15] Newman describes an algorithm based on a characteristic of clustering quality called *modularity*, a measure that takes

---

* This work has been supported by the Department of Energy under contract W-705-ENG-36.

W. Aiello et al. (Eds.): WAW 2006, LNCS 4936, pp. 117–128, 2008.

into account the number of in-cluster edges and the expected number of such edges. (We formally define and discuss modularity in more detail in the next section.) A faster version of the algorithm from [15] was described by Clauset *et al.* in [6]. Several algorithms have been proposed based on the eigenvectors of the graph Laplacian, e.g., [19], [16]. In all previous cases the algorithms reported in the literature are either not fast enough, or are inaccurate.

In this paper we will describe a new approach for GC that uses our newly discovered relationship between the GC and the minimum weighted cut problems. The *minimum weighted cut* (MWC) problem is, given a graph $G = (V, E)$ with real weights on its edges, find a partition of V such that the set of all edges of G that join vertices from different sets of the partition, called a *cut* of the partition, is of minimum weight. GC looks related to the MWC problems since, in a good quality clustering, the weight of the edges between different sets of the partition (the cut) should be small compared to the weight of the edges inside the sets. But the MWC problem can not be directly applied to solve the GC problem since it does not take into account the sizes of the subgraphs induced by the cut (e.g., it is likely that the minimum cut will consist of the edges incident to a single vertex). There are some minimum cut based clustering algorithms, e.g., [9], that use maximum flow computations combined with heuristics, but they are typically slower than modularity based algorithms, e.g. [6], and, moreover, they cannot determine the optimal number of clusters and, instead, construct a hierarchical decomposition of the set of all vertices of the graph.

In this paper we prove that the problem of finding a partition of a graph G that maximizes the modularity can be reduced to the problem of finding a MWC of a weighted complete graph on the same set of vertices as G. We then show that the resulting minimum cut problem can be solved by modifying existing fast algorithms for graph partitioning. We demonstrate by experiments that our algorithm has generally a better quality and is much faster than the best existing GC algorithms.

2   Our Clustering Algorithm

2.1   Graph Clustering as a Minimum Cut Problem

As there is no formal definition of clustering and what the clusters of a given graph are, in general it is not possible to determine if a certain partition is the "correct" clustering or which of two alternative partitions of a graph corresponds to a better clustering. For that reason, researchers have used their intuition to define measures for cluster quality that can be used for comparing different partitions of the same graph. One such measure, introduced in [18,17], which has received considerable attention recently, is the *modularity* of a graph. Given an n-vertex m-edge graph $G = (V(G), E(G))$ and a partition $\mathcal{P}$ of $V(G)$ into k subsets (clusters) $V_1, \ldots, V_k$, the modularity $Q(\mathcal{P})$ of $\mathcal{P}$ is a number defined as

$$Q(\mathcal{P}) = \frac{1}{m} \sum_{i=1}^{k} (|E(V_i)| - \mathrm{Ex}(V_i, \mathcal{G})),$$

where $E(V_i)$ is the set of all edges of G with endpoints in V_i and $\text{Ex}(V_i, \mathcal{G})$ is the expected number of such edges in a random graph with a vertex set V_i from a given random graph distribution $\mathcal{G}$. $Q(\mathcal{P})$ measures the difference between the number of in-cluster edges and the expected value of that number in a random (e.g., without cluster structure) graph on the same vertex set. Larger values of $Q(\mathcal{P})$ correspond to better clusterings.

Having the definition of $Q(\mathcal{P})$, we can formulate the clustering problem as finding a partition $\mathcal{P} = \{V_1 \cup \ldots \cup V_k\}$ of $V(G)$ such that

$$\sum_{i=1}^{k}(|E(V_i)| - \text{Ex}(V_i, \mathcal{G})) \to \max. \tag{1}$$

Clearly

$$\max_{\mathcal{P}}\{ \sum_{i=1}^{k}(|E(V_i)| - \text{Ex}(V_i, \mathcal{G}))\}$$

$$= -\min_{\mathcal{P}}\{ -\sum_{i=1}^{k}(|E(V_i)| - \text{Ex}(V_i, \mathcal{G}))\}$$

$$= -\min_{\mathcal{P}}\{ (|E(G)| - \sum_{i=1}^{k} |E(V_i)|) - (|E(G)| - \sum_{i=1}^{k} \text{Ex}(V_i, \mathcal{G}))\}$$

$$= -\min_{\mathcal{P}}\{ |\text{Cut}(\mathcal{P})| - \text{ExCut}(\mathcal{P}, \mathcal{G})\},$$

where $\text{Cut}(\mathcal{P})$ is defined as the cut of $\mathcal{P}$ and $\text{ExCut}(\mathcal{P}, \mathcal{G})$ the expected value of $\text{Cut}(\mathcal{P})$ for a random graph from $\mathcal{G}$.

Hence, instead of problem (1), one can address the problem of finding a partition $\mathcal{P}$ of G such that

$$|\text{Cut}(\mathcal{P})| - \text{ExCut}(\mathcal{P}, \mathcal{G}) \to \min. \tag{2}$$

The last expression shows that we can solve (1) as a problem of finding a MWC in a complete graph G' with a vertex set $V(G)$ and weight $\text{weight}(i, j)$ on any edge $(i, j) \in E(G')$ defined by

$$\text{weight}(i, j) = \begin{cases} 1 - p_{ij}, & \text{if } (i, j) \in E(G) \\ -p_{ij}, & \text{if } (i, j) \notin E(G), \end{cases} \tag{3}$$

where p_{ij} is the probability that there is an edge between vertices i and j in a random graph from the class $\mathcal{G}$. Then, problem (1) is equivalent to the problem of finding a partition $\mathcal{P}'$ of G' such that

$$|\text{Cut}(\mathcal{P}')| \to \min. \tag{4}$$

We summarize these observations in the following theorem.

Theorem 1. *The problem of finding a partition of a given graph* $G = (V, E)$ *that minimizes the modularity can be reduced in* $O(|V| + |E|)$ *time to the problem of finding a minimum weight cut in a complete graph* $G' = (V, E')$ *with edge weights given by (3).*

For the reduction time bound in Theorem 1 we assume that the edges of $E' \setminus E$ are defined implicitly. There are several choices for $\mathcal{G}$ that have been favored by various researchers. The random graph model $G(n, p)$ of Erdös-Renyi [7] defines n vertices and puts an edge between each pair with probability p. Clearly, the expected number of edges of $G(n, p)$ is $\binom{n}{2}p$. Hence, for a graph with expected number of edges m

$$p_{ij} = p = \frac{m}{\binom{n}{2}} \ . \tag{5}$$

One disadvantage of the $G(n, p)$ model is that it fails to capture important features of the real-world networks, in particular, the degree distribution. As has been recently observed [3], many important types of networks like technological networks (the Internet, the WWW), social networks (collaboration networks, online social networks), biological networks (protein interactions) have degree distributions that follow a *power law*, e.g., the fraction of the vertices that have degree $k > 0$ is roughly proportional to $\alpha k^{-\lambda}$ for some constants α and $\lambda > 0$. Such networks are called *scale-free*. In comparison, the degrees of a random graph from the $G(n, p)$ model follow a Poisson distribution, i.e., the probability that a given vertex has degree k is $\binom{n}{k}p^k(1 - p)^{n-k}$ and the expected degree of each vertex is pn. Hence, the Erdös-Renyi model may not be suitable as a choice for $\mathcal{G}$ when used for determining the community structure of graphs of the above type.

One model that takes into account the degrees of the vertices is studied by Chung and Lu in [5]. In that model, the probability that there is an edge between a vertex i and a vertex j is

$$p_{ij} = \frac{d_i d_j}{\sum_{k=1}^{n} d_k} \ , \tag{6}$$

where $d_1, \cdots, d_n$ are positive reals corresponding to the degrees of the vertices such that $\max_{1 \leq i \leq n} d_i^2 < \sum_{i=1}^{n} d_i$. (The last condition guarantees that such a graph exists if all numbers d_i are integers.) We will refer to that model as the Chung-Lu (CL) model. Clearly, in the CL model, the expected degree of vertex i is d_i, compared with pn (i.e., independent on i) in the $G(n, p)$ model.

In the next section we will describe an efficient method for finding a MWC of a graph G' with weights on the edges satisfying (3) and p_{ij} defined by (5) or (6).

2.2   Finding a MWC Using Multilevel Graph Partitioning

Above we established an important relationship between the graph clustering and the MWC problems, i.e., that the problem of finding a partition of a given graph that maximizes the modularity can be reduced to the problem of finding a minimum weight cut. Most existing work on the MWC problem considers the case where all weights are non-negative. The MWC problem in the case of non-negative weights is known to be polynomially solvable, e.g., by using algorithms for computing maximum flows [1]. In contrast, the MWC problem in case of real-value weights is NP-hard and there is very little known for the general version of the problem. Here we show that available heuristics for another related problem, graph partitioning, can be adapted to solve this version of the MWC problem.

Overview of the multilevel partitioning method. Formally, the *graph partitioning* (GP) problem is, given a graph $G = (V, E)$, find a partition (V_1, V_2) of V such that $||V_1| - |V_2|| \leq 1$ (i.e., the partition is *balanced*) and $\text{Cut}(V_1, V_2)$ is minimized. (Some versions of the problem consider partitions of arbitrary cardinalities.) Note that, in comparison with the minimum cut problem, there is the additional requirement for a balanced partition. Because of its important applications, e.g., in high performance computing and VLSI design, GP is a well-researched problem for which very efficient methods have been developed. One such approach is the multilevel GP, which is both fast and accurate for a wide class of graphs that appear in practical applications. Inspired by the multigrid method from computational mathematics, it has been used in the works of Barnard and Simon [4], Hendrickson and Leland [10], Karypis and Kumar [12,13], and others. The method for bisecting a graph consists of the following three phases(Figure 2.2):

Coarsening phase. The original graph G is coarsened by partitioning it into connected subgraphs and replacing each of the subgraphs by a single vertex and replacing the set of the edges between any pair of shrunk subgraphs by a single edge. Moreover, a weight of each new vertex (respectively edge) is assigned equal to the sum of the weights of the vertices (respectively edges) that it represents. (Weights on the original vertices of G will be defined depending on whether the $G(n, p)$ or the CL model has been used, as detailed below.) The resulting graph is coarsened repeatedly by the same procedure until one gets a graph of a sufficiently small size. Let $G_0 = G, G_1, \ldots, G_l$ be the resulting graph sequence.

Partitioning phase. The graph G_l is partitioned into two parts using any available partitioning method (e.g., spectral partitioning or the Kernighan-Lin (KL) algorithm [14]).

Uncoarsening and refinement phase. The partition of G_l is projected on G_{l-1}. Since the weight of each vertex of G_l is a sum of the weights of the corresponding vertices of G_{l-1}, then the partition of G_{l-1} will be balanced if the partition of G_l is and the cut of both partitions will have the same weight. However, since G_{l-1} has more vertices than G_l, it has more degrees of freedom and, therefore, it is possible to refine the partition of G_{l-1} in order to reduce its cut size. For this end, the projection of the partition of G_l is followed by a refinement phase, which is usually based on the KL algorithm. In the same way, the resulted partition of G_{l-1} is converted into a partition of G_{l-2} and refined, and so on until a partition of G_0 is found.

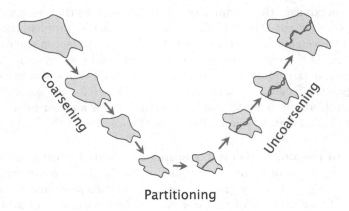

Partitioning

Fig. 1. The stages of multilevel partitioning

Kernighan-Lin refinement. Since the refinement step is the most involved part of the algorithm, and which ultimately determines its accuracy and efficiency, we will describe it in more detail. It has been shown [13] that the KL algorithm can be a good choice for performing the refinement.

The KL algorithms involves several iterations, each consisting of moving a vertex from one set of the partition to the other. Let $\mathcal{P} = \{P_1, P_2\}$ be the current partition. For each vertex u of the graph a *gain* for u is defined as

$$gain(u) = \sum_{v \in N(u) \backslash P(u)} weight(u, v) - \sum_{v \in N(u) \cap P(u)} weight(u, v), \qquad (7)$$

where $N(u)$ is the set of all neighbors of u and $P(u)$ is that set of $\mathcal{P}$ that contains u. $gain(u)$ measures how the weight of the cut will be affected if u is moved from $P(u)$ to the other set of $\mathcal{P}$. The KL algorithm then selects a vertex w from the smaller set of the partition with a maximum gain, moves it to the other set, and updates the gains of the vertices adjacent to w. Moreover, w is marked so that it will not be moved again during that refinement step. The process is continued until either all vertices have been moved, or the 50 most recent moves have not led to a better partition. At the end of the refinement step, the last $s \leq 50$ moves that have not improved the partition are reversed.

Implementation. The implementation of our algorithm for clustering is based on the version of multilevel partitioning implemented by Karypis and Kumar [12,13], which has been made freely available as a software package under the name METIS. Note that graph partitioning, minimum cut, and clustering are related, but with important differences problems, as illustrated in Table 1. We already showed how the clustering problem can be reduced to a minimum cut problem and here we will show how the resulting minimum cut problem can be solved by a graph partitioning algorithm based on METIS. Because of the differences between graph partitioning and MWC, we have to make some evident changes. For instance, since graph partitioning requires balanced partitions,

Table 1. Comparison between the clustering, minimum cut, and partitioning problems

Problem	Clustering	Minimum Cut	Graph Partitioning
Objective	Minimize modularity	Minimize cut size	Minimize cut size
Balance of partition	Sizes may differ	Sizes may differ	Equal sizes
Cardinality of partition	To be computed	To be computed	An input parameter

we have to drop the requirement for balance of the partition. We have also to determine the cardinality of the partition that minimizes the cut size. But the main implementation difficulty is related to the size of G'. Although the original graph, G, is typically sparse, i.e., has n vertices and $O(n)$ edges, the transformed one, G', is always dense, as it has $\binom{n}{2} = \Omega(n^2)$ edges. The main challenge will be to construct an algorithm whose complexity is close to linear on the size of the original graph, rather than on the size of the transformed one. We have shown that it is possible to simulate an execution of the multilevel algorithm on G' by explicitly maintaining information only about the edges from the original graph G and implicitly taking into account the remaining edges by modifying the formulae for computing weights and gains. For instance, if $\mathcal{P} = \{P_1, P_2\}$ is a partition of $V(G)$ and we have computed the value of the cut $\mathrm{cut}(P_1, P_2)$ of G corresponding to $\mathcal{P}$ and maintain the values of $n_1 = |P_1|$ and $n_2 = |P_2|$, then the cut in G' corresponding to $\mathcal{P}$ is

$$\mathrm{cut}(P_1, P_2) - n_1 n_2 p$$

in the case of the $G(n, p)$ model and hence can be computed in $O(1)$ time. A similar formula holds for the case of the CL model.

Clustering into an optimal number of clusters. The algorithm described above is a *bisection* algorithm, i.e., it finds a partition (and hence clustering) of the input graph into two parts. Our algorithm for an arbitrary number of clusters uses the following recursive procedure. We run the bisection algorithm described above and let $\mathcal{P}$ be the resulting partition. If $\mathcal{P}$ consists of only one set (i.e., the original graph G does not have a good cluster partition), we are done. Else, we run recursively the bisection algorithm on the two subgraphs G_1 and G_2 of G induced by the vertices of the two sets of $\mathcal{P}$. It is important to keep, during that recursive call, the weights of the edges computed during the first iteration instead of recomputing them based on G_1 and G_2. The reason is that the random graph model based on G will be different than those based on G_1 and G_2 since formulae (5) and (6) will produce different values for p_{ij}. It can be proven that, if the bisection algorithm finds a minimum bisection cut, the recursive algorithm described above finds a minimum cut (of any number of parts) and hence finds a clustering maximizing the modularity.

Time analysis. By using the analysis of Fiduccia and Mattheyses of the KL algorithm from [8], it follows that clustering any network of n vertices and m edges into two communities by our algorithm takes $O(n \log n + m)$ time, where

n and m are the numbers of the nodes and links of the network, respectively. Finding a clustering in optimal number of k parts takes $O((n \log n + m)d)$ time, where d is the depth of the dendrogram describing the clustering hierarchy. Although the worst-case value of d can be $\Omega(k)$, typically $d = O(\log k)$ [6].

3   Experiments

We performed a number of experiments on randomly generated graphs in order to measure the accuracy of our algorithm and its efficiency as well as to compare it with previous algorithms. We chose Newman-Girvan algorithm [18] and Clauset-Newman-Moore algorithm [6] since they are considered one of the best existing algorithms and because of the code availability.

3.1   Comparison with Newman-Girvan Algorithm

Following the experimental setting of [18], we generated random graphs with 128 vertices and 4 communities of size 32 each. The expected degree of any vertex is 16, but the *outdegree* (the expected number of neighbors of a vertex that belong to a different community) is set to i in the i-th experiment ($i \leq 16$). Hence, higher values of i correspond to graphs with weaker cluster structures. The experiment is intended to measure the sensitivity of the algorithm to the quality of clustering.

Table 2. Comparing the quality of the clustering of our algorithm and [18]

Outdegree	Degree	Newman-Girvan	Ours
1	16	1.00	1.00
2	16	1.00	1.00
3	16	0.98	0.99
4	16	0.97	0.99
5	16	0.95	0.99
6	16	0.85	0.97
7	16	0.60	0.91
8	16	0.30	0.70

Table 2 compares the quality of the clusterings produced by Newman-Girvan's algorithm and ours. A clustering produced by any of the algorithms is considered "correct" if it matches the original partition of communities from the graph generation phase. (Note that, due to the probabilistic nature of the graphs, the clustering that maximizes the modularity might be different from the original partition, especially if the modularity is low.)

Our algorithm classifies correctly more than 99% of the edges for outdegrees $0, 1, 2, 3, 4, 5$ and in all cases it is better than Newman-Girvan's (more than twice better for the case $i = 8$).

3.2   Comparison with Clauset-Newman-Moore Algorithm

Table 3 compares the performance of our algorithm with Clauset, Newman, and Moore's algorithm [6]. That algorithm has the same quality of the clustering as [15], but is claimed to be much faster. The test graphs in all experiments are random graphs with different number of clusters, sizes, densities, and modularities. Each experiment has been run 100 times on different random graphs.

Table 3. Comparison between the performances of our algorithm and [6]. Q_{orig} is the modularity of the partition used during graph generation, "$Q_{CNM} >$", "$Q_{ours} >$", and "$Q =$" are the percentages of the cases where the algorithm [6] produced a better modularity, our algorithm produced a better modularity, or both algorithms produced equal modularities, respectively. T_{CNM} and T_{ours} are the times of the algorithm from [6] and ours, respectively.

Exp.	# vert.	# edges	# clust.	Q_{orig}	$Q_{CNM} >$	$Q_{ours} >$	$Q =$	T_{CNM}	T_{ours}
1	200	8930	2	.388	0	8	92	.61	.03
2	300	14891	3	.466	0	22	78	1.01	.05
3	400	21853	4	.474	0	42	58	1.24	.11
4	500	29801	5	.463	0	57	43	1.71	.23
5	600	38776	6	.446	0	70	30	2.25	.15
6	700	48706	7	.426	1	87	12	2.90	.22
7	800	59666	8	.406	2	96	2	3.71	.33
8	900	71546	9	.387	1	99	0	4.44	.35
9	200	9932	2	.298	0	8	92	.68	.04
10	200	4967	2	.299	0	27	73	.54	.03
11	200	2458	2	.298	0	50	50	.61	.02
12	200	1238	2	.295	6	92	2	.46	.00
13	400	41856	4	.176	32	63	5	1.61	.18
14	400	43607	4	.154	39	60	1	1.66	.10
15	400	47797	4	.122	89	11	0	1.84	.07
16	400	8537	4	.244	0	100	0	1.35	.02
17	400	4879	4	.273	0	100	0	1.33	.01
18	400	2653	4	.308	0	100	0	1.33	.03
19	400	1449	4	.370	0	100	0	1.36	.04
20	400	888	4	.375	0	100	0	1.35	.02
21	400	629	4	.394	0	100	0	1.34	.03

In experiments 1–15 the random graphs were generated in the following way: a graph with no edges is created whose vertices are divided into subsets that correspond to the clusters; then edges are created with probability p_{in} between vertices in the same subset and with probability p_{out} between vertices from different subsets. Experiments 1–8 compare how the performance of the algorithms depends on the number of clusters, which vary from 2 to 9. The results indicate that our algorithm produces always clusterings with better quality, and the difference increases when the number of the clusters grows. In experiments 9–12 the test graphs have the same number of vertices, number of cluster, and modularity, but different densities. Those experiments show that our algorithm is more sensitive when the density decreases, and in all the cases our algorithm performs better. In experiments 13–15, we compare the algorithms when the modularity (the quality of the original clustering) is very low. We determined that with modularity less than approximately 0.15 the algorithm from [6] is better, and if the modularity is greater than 0.15 our algorithm is better. In all the above experiments, the running time of our algorithm is considerably smaller, whereby our algorithm is between 7 and 30 times faster than the algorithm from [6].

Finally, in experiments 16–21 the random graphs were created such that their expected degree sequences satisfy a power law distribution. The exponent of the density function varies from -1.0 in experiment 16 to -2.0 in experiment 21 in increments of -0.2. The results of the experiments imply that in the case of power-law degree distributions (scale-free graphs) the quality of our algorithm consistently beats the one of the algorithm from [6], while our time is in average 54 times smaller than theirs.

3.3   Testing on Real-World Data Graphs

We tested our algorithms on a number of real-world graphs such as the *nd.edu* domain data [2], the United States college football data [11], and the Zachary's karate club network [20]. In all cases our algorithm produced clustering consistent with our previous knowledge of the communities. For example, we describe in more detail here the Zachary club network. This example is a standard benchmark for community detection algorithms, describing the interactions between the members of a karate club, which consequently split into two because of between the members, thereby revealing the hidden communities of the original network. As shown on Figure 2, our algorithm classified correctly the members

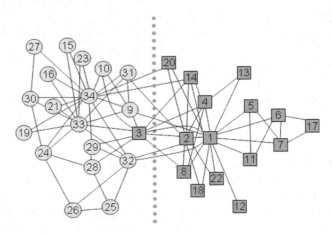

Fig. 2. Zachary's karate club network. Members of the communities resulting after the split are denoted by circles and squares, respectively. The communities found by our algorithm are separated by the vertical line.

Table 4. Measuring the scalability of our algorithm. p_{in} (respectively p_{out}) is the expected fraction of the number of in-cluster (respectively between cluster) edges to the number of all pairs of vertices from the same set (respectively different sets) of the partition used for graph generation.

p_{in}	p_{out}	Vertices	Edges	Total size	Time (sec.)
0.10	0.01	5,000	406,125	411,125	1.77
0.14	0.01	6,000	764,126	770,126	3.09
0.18	0.01	7,000	1,283,398	1,300,398	3.22
0.20	0.011	8,000	1,863,710	1,871,710	6.66
0.20	0.013	9,000	2,418,730	2,427,730	5.68
0.21	0.014	10,000	3,153,106	3,163,106	7.27
0.22	0.015	15,000	6,295,801	6,310,801	15.18

of the two subgroups, except for node 10. That node has the same number of links (five) to both communities, hence adding it to the smaller community results in a greater modularity (e.g., our partitioning has a better modularity than the "real" one.)

3.4   Measuring the Scalability

We also tested the speed of our algorithms by running them on a 2 GHz desktop computer on graphs of different sizes. The results are illustrated on Table 4 and clearly show the extraordinary speed and scalability of our algorithms.

4   Conclusion

This paper proposes a new approach for graph clustering by reducing the clustering problem to a minimum cut problem and then solving the latter problem by applying methods for graph partitioning. Our proof-of-concept implementation, based on the METIS partitioning package, demonstrated the practicality of the approach. The changes we made to METIS were minimal and various improvements and refinements that take into account the specifics of the clustering problem, use alternative minimum cut or graph partitioning algorithms, or apply heuristics and parameter adjustments in order to improve the accuracy are possible and will be topics of further research.

Acknowledgement. The author is indebted to Melih Onus for helping with the programming and most of the experiments and for many helpful discussions. We also would like to thank the developers of METIS for making their source code publicly available.

References

1. Ahuja, R.K., Magnanti, T.L., Orlin, J.B.: Network Flows: Theory, Algorithms, and Applications. Prentice-Hall, Englewood Cliffs (1993)
2. Albert, R., Jeong, H., Barabási, A.L.: Diameter of the World Wide Web. Nature 401, 130 (1999)

3. Barabási, A.L., Albert, R.: Emergence of Scaling in Random Networks. Science 286, 509–512 (1999)
4. Barnard, S.T., Simon, H.D.: A fast multilevel implementation of recursive spectral bisection for partitioning unstructured problems. Concurrency: Practice and Experience 6, 101–107 (1994)
5. Chung, F., Lu, L.: Connected components in random graphs with given degree sequences. Annals of Combinatorics 6, 125–145 (2002)
6. Clauset, A., Newman, M., Moore, C.: Finding community structure in very large networks. Phys. Rev. E 70, 066111 (2004)
7. Erdos, P., Renyi, A.: On random graphs. Publicationes Mathematicae 6, 290–297 (1959)
8. Fiduccia, C.M., Mattheyses, R.M.: A linear time heuristic for improving network partitions. IEEE Design Automation Conference, 175–181 (1982)
9. Flake, G.W., Tarjan, R.E., Tsioutsiouliklis, K.: Graph Clustering and Minimum Cut Trees. Internet Mathematics 1, 385–408 (2004)
10. Hendrickson, B., Leland, R.: A Multilevel Algorithm for Partitioning Graphs. In: ACM/IEEE conference on Supercomputing (1995)
11. Girvan, M., Newman, M.: Community structure in social and biological networks. Proc. Natl. Acad. Sci. USA 99, 7821–7826 (2002)
12. Karypis, G., Kumar, V.: Multilevel graph partitioning schemes. In: International Conference on Parallel Processing, pp. 113–122 (1995)
13. Karypis, G., Kumar, V.: A fast and high quality multilevel scheme for partitioning irregular graphs. SIAM Journal on Scientific Computing 20(1), 359–392 (1999)
14. Kerninghan, B.W., Lin, S.: An efficient heuristic procedure for partitioning graphs. The Bell System Technical Journal (1970)
15. Newman, M.: Fast algorithm for detecting community structure in networks, Phys. Phys. Rev. E 69, 066133 (2004)
16. Newman, M.: Finding community structure in networks using the eigenvectors of matrices. Phys. Rev. E 74, 036104 (2006)
17. Newman, M.: Mixing patterns in networks. Phys. Rev. E 67, 026126 (2003)
18. Newman, M., Girvan, M.: Finding and evaluating community structure in networks. Phys. Rev. E 69, 026113 (2004)
19. White, S., Smyth, P.: A Spectral Clustering Approach to Finding Communities in Graphs. In: Proceedings of the SIAM International Conference on Data Mining (2005)
20. Zachary, W.W.: An information flow model for conflict and fission in small groups. Journal of Anthropological Research 33, 452–473 (1977)

A Phrase Recommendation Algorithm Based on Query Stream Mining in Web Search Engines

M. Barouni-Ebrahimi and Ali A. Ghorbani

Faculty of Computer Science, University of New Brunswick, Fredericton, Canada
{m.barouni,ghorbani}@unb.ca

Abstract. In this paper, a phrase recommender algorithm is proposed that suggests the related frequent phrases to an incomplete user query. The suggested phrases are extracted from past user queries based on the frequency rate of the phrases. A query recommender algorithm called OQD (Online Query Discovery) has also been designed for comparison purposes. Simulation results show the efficiency of the proposed phrase recommender algorithm compared to the OQD. The phrase recommender algorithm significantly reduces the size of the candidate set, which results in smaller memory usage and better performance, while recommending more appropriate phrases to the user.

1 Introduction

User queries submitted to the web search engines are not always informative enough for retrieving the related pages to the user intention. The main problem is that users may not know the best query items they should enter to get the most related web pages to their intentions. They may not be familiar with the specific keywords in that domain knowledge. A user may remember only a part of the phrase that he/she wants to use in the query string. Sometimes the user does not know how to order the keywords (most web search engines are sensitive to the order of the keywords) or even does not know the correct spelling of a specific keyword in the query string. A novice user sometimes sends an imperfect query and scans the returned web pages (even reads a number of the returned documents) to prepare a more precise query by finding new related keywords in the documents. Although this partially treats the problem, it is often not a straight forward task. Finding related keywords in a list of web pages full of unrelated information is a frustrating task for the impatient users.

Significant efforts have been put into designing appropriate mining techniques for web search query logs for query recommendation as well as query expansion. In the case of query recommendation, suggesting related queries is considered, while appending the related items to a newly submitted query is the subject of query expansion. Google has launched the google suggestion service that recommends relevant terms for query completion. The algorithm has not been published; therefore, the efficiency of the algorithm cannot be evaluated. A deficiency of Google's method is that the suggestions are only provided for the first phrase in a query. A Phrase is a sequence of words in a query (it can be a word or the

W. Aiello et al. (Eds.): WAW 2006, LNCS 4936, pp. 129–136, 2008.

whole multi-item query) that is frequently used by users. If the user attempts to enter two different frequent phrases in the query, Google does not provide any suggestions for the second one. It seems that Google deals with the whole query as one phrase. On the other hand, the suggestions are only the words that can be appended to the end of the query. What if the user knows only the middle part of a query?

A query completion algorithm has been proposed in [3] that suggests the frequent words of the last incomplete word in the query. The suggested words are extracted from the documents that contain the previous words of the incomplete query. The differences between the document space and the query space have been emphasized in [6] due to the fact that the similarities, preferences and terms in document space are not necessarily compatible with the ones in the query space. We therefore prefer to extract the frequent phrases from the previous queries rather than the related documents which is employed in [3]. Our algorithm suggests the complementary phrases for an incomplete user query. The complementary phrases are the ones that contain a part of the uncompleted user query. The phrases are actually the conceptual frequent phrases mined from past users' queries. The conceptual frequency rate is a new definition introduced in this paper, which we believe is more appropriate for frequent phrase extraction from query streams compared to similar definitions.

The rest of the paper is organized as follows. In the next section, the conceptual frequency rate definition for mining frequent sequences in a data stream is described. In Sections 3 and 4, the phrase recommender algorithm is given, and the simulation results are provided. The paper is finally concluded in Section 5.

2   Conceptual Frequency Rate

Legacy algorithms for frequent sequence mining extract the frequent items from a static dataset. A dataset is a set of transactions. Each transaction contains a list of items. The algorithms find the maximal frequent sequences of items that satisfy a user defined minimum support value. Huge number of daily queries has changed the nature of the query logs to query streams. The differences between mining data streams and mining static datasets have been pointed in [5]. Firstly, each transaction within a data stream should at most be examined once. Secondly, a mechanism is needed to bind the number of candidate elements due to the continuous generation of new data elements. Finally, the mining result should be available whenever it is requested regardless of the stage of the process. Charikar et al. [4] have proposed a one-pass algorithm for estimating the most frequent items in a data stream. The algorithm addresses the problem that comes up in the context of search engines, where there is a query stream sent to the search engine and the goal is to find the most frequent queries handled in some period of time. A query is an item in the algorithm, while in this paper; we are interested in finding the common phrases used in the user queries. A query is composed of a number of phrases. Each phrase is a word (generally an item) or a sequence of words.

Min-support is not a suitable threshold in data stream mining due to the fact that there is an unlimited number of transactions in a data stream. A phrase may be frequently seen in some periods of time, while it does not frequently occur in other periods. Therefore, the phrase would be a frequent phrase only in those periods of time. We proposed frequency rate in [2], a new definition for mining frequent sequences in data streams, which is defined as follows:

$$f_P = \frac{n_P}{t_c - t_P + 1}, \tag{1}$$

where f_P is the frequency rate of the phrase P, n_P is the occurrence number of P, t_c is the current transaction number and t_P is the birth number of P. Two parameters t_c and t_P are actually the current date and the phrase birth date respectively. The frequency rate, f_P, is a real number between 0 and 1. A frequency rate of zero for a phrase means that the respective phrase has not been seen in any of the transactions in the data stream, while a frequency rate of one implies that the phrase is present in all of the transactions in the data stream. A phrase is frequent if its frequency rate satisfies the user defined frequency rate, f_u.

Our proposed algorithm in [2] extracts the frequent phrases of a given data stream. The algorithm satisfies the data stream mining constraints mentioned earlier. It examines a transaction only once. It also decreases the number of elements in the candidate set compared to the other general algorithms. The reason is that the algorithm does not add a longer phrase to the candidate set, unless the shorter sequences of the phrase are frequent. The frequent phrases can also be extracted from the candidate set regardless of the stage of the algorithm. The algorithm creates a phrase in the first visit, monitors its frequency rate while reading the data stream. It extracts the phrase as a frequent phrase as long as it is frequently seen in the transactions and deletes it as soon as its frequency rate does not satisfy f_u. The phrase may be born again in another part of the data stream.

As an example, consider the data stream DS in which transactions $T_1 = A_1 A_2 A_4 A_5$, $T_2 = A_2 A_4$, $T_3 = A_1 A_2 A_3 A_5$ and $T_4 = A_2 A_3$ are repeated periodically. The frequent phrases resulting from different f_u values are shown in Table 1.

Table 1. Frequent Phrases and Conceptual Frequent Phrases of the sample Data Stream DS

f_u	Frequent phrases ($f_P > f_u$)	Conceptual frequent phrases ($cf_P > f_u$)
0.2	A_1, $A_1 A_2$, $A_1 A_2 A_3$, $A_1 A_2 A_3 A_5$, $A_1 A_2 A_4$, $A_1 A_2 A_4 A_5$, A_2, $A_2 A_3$, $A_2 A_3 A_5$, $A_2 A_4$, $A_2 A_4 A_5$, A_3, $A_3 A_5$, A_4, $A_4 A_5$, A_5	$A_1 A_2 A_3 A_5$, $A_1 A_2 A_4 A_5$, $A_2 A_3$, $A_2 A_4$
0.4	A_1, $A_1 A_2$, A_2, $A_2 A_3$, $A_2 A_4$, A_3, A_4, A_5	$A_1 A_2$, $A_2 A_3$, $A_2 A_4$, A_5
0.6	A_2	A_2

The frequency rate definition is suitable for data stream mining, however, it is still not appropriate for the real world cases such as phrase recommendation in web search engines. The phrase recommender algorithm needs to mine the longest frequent phrases that the users frequently submit in their queries. A way for mining the longest frequent patterns is to find the maximal frequent sequences. There are cases where a frequent phrase is also a part of longer frequent phrases. Such a phrase is frequent regardless of its occurrences in the longer ones. The problem of maximal frequent phrase mining techniques is that they do not classify such phrase as frequent. Consider the phrase "New York city map" as a frequent phrase within user queries such as "New York city map with streets" and "New York city map Manhattan". On the other hand, consider the phrase "city map" as a frequent phrase in user queries such as "London city map" and "Toronto city map". A frequent sequence miner finds all the sub phrases of these two sequences as a frequent phrase. This is not the proper answer due to the fact that phrases like "York city", "York city map" are not the common query phrases. On the other hand, a maximal frequent sequence miner only finds "New York city map" as a frequent phrase. The phrase "city map" is a part of the bigger phrase and therefore is not recognized as an independent frequent phrase.

Considering the sample data stream DS, there are 4 transactions (T_1, T_2, T_3 and T_4) that should be chosen as frequent phrases when f_u=0.2. A frequent phrase miner finds 16 frequent phrases for f_u=0.2, while a maximal frequent phrase miner finds only two phrases T_1 and T_3. None of these results are actually the proper answers. Conceptual frequency rate is defined as follows:

$$cf_P = \frac{cn_P}{t_c - t_P + 1},$$ (2)

where cf_P represents the conceptual frequency rate of the phrase P, cn_P denotes the conceptual occurrence number of P, which is defined by Equation 3, t_c and t_P represent the current transaction number and the phrase birth number, respectively.

$$cn_P = n_P - \sum_{P_i \in SS_P} cn_{P_i},$$ (3)

where SS_P is a set of all frequent phrases such as P_i, such that P is a part of P_i.

Equation 2 implies that a phrase is conceptually frequent only if it is frequent regardless of its occurrence number in the longer frequent phrases. For the sample data stream DS, the frequent phrases resulting from different f_u values based on conceptual frequency rate definition are shown in Table 1. All four transactions are chosen as frequent phrases for f_u=0.2. The phrases "$A_1 A_2$", "$A_2 A_3$", "$A_2 A_4$" and "A_5" are conceptually frequent for f_u=0.4. From another perspective, these phrases are the longest frequent phrases with frequency rates higher than 0.4 and if they are removed from data stream, there are no more phrases with a frequency rate higher than 0.4.

The proposed OFSD algorithm in [1], extracts the conceptual frequent phrases from a given data stream. In the method, candidate set (CS) is a directed graph.

Each node of the CS is a candidate phrase and an edge $l(i,j)$ is considered to be directed from P_i to P_j, where $T : P_i \rightarrow P_j$ (implies P_j immediately follows P_i in the transaction). The detail of the algorithm along with the proof of correctness can be found in [1].

3 Phrase Recommender Algorithm

By applying OFSD algorithm to the user queries in a web search engine, the frequent phrases of the queries are collected continuously in a dataset. The frequent phrases used by previous web surfers are suitable resources for recommending relevant phrases to a new user. An important issue in this context is that the frequency rates of the phrases may change with time. This is due to the fact that some of the users' interests may not last in for a long term. There are certain events (e.g. soccer world cup), which result short-time frequent phrases in the users' queries. The algorithm is able to extract a short-time frequent phrase in its related queries list and then will forget it whenever it is not frequent anymore. A phrase has a chance to be born in different query lists, be selected as a frequent phrase and die after a while. A phrase is considered for recommendation as long as it is frequently used by the users.

When a user enters a query segment in the user interface of web search engine, the phrase recommender algorithm searches to find the conceptual frequent phrases that contain the query segment. This list of phrases is called relevant list. The relevant list is recommended to the user due to the fact that the phrases within the relevant list have been previously frequently used by other users. This helps the user in different aspects. Firstly, the user may find new related keywords to its query in the relevant list. Secondly, he/she may realize the correct sequence of the query items he/she wants to enter. The sequence of items in a multi-element phrase is important because the common web search engines (e.g. Google) are sensitive to the sequence of the items in the query. Finally, he/she may recognize the syntax of the phrase items. A rational assumption here is that the relevant list consists of well-formed phrases (meaningful phrases without any syntax errors). The fact is that the probability of entering a well-formed phrase is more than entering any certain incorrect form of the phrase, therefore, the frequency rate of a well-formed phrase in a query stream is more than any specific incorrect form of that phrase.

Relevant list of phrases for recommendation is initially constructed based on the last uncompleted word of the incomplete query. If the size of the relevant list is more than a threshold (LS), the last two or more sequence of items in the incomplete query will be used for frequent phrase selection to decrease the size of the relevant list. The threshold LS is set to 10 in the Google suggest website. The reason is that it is not easy for the user to go through all the phrases in a long relevant list. For the cases that the size of the relevant list is still greater than LS, even by using all the items in the query segment, the algorithm only returns the top LS phrases as the relevant list. The relevant list is ordered based on the frequency of the phrases. The most frequent phrase is on top of the relevant list.

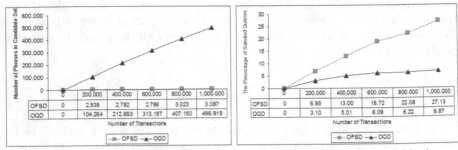

a: Number of All Phrases in the Candidate Set   b: The Percentage of the Satisfied Queries

Fig. 1. The comparison between the phrase recommender algorithm based on OFSD algorithm and the OQD algorithm for 1,000,000 queries, f_u=0.0008

4   Simulation Results

We have designed a query recommender algorithm called OQD (Online Query Discovery), which is used for comparison purposes. A query is a non-separable item in the majority of studies done by other researchers. Here, we show the efficiency of *phrase recommendation* instead of *query recommendation*. The OQD keeps track of the number of query occurrences. The algorithm is similar to the phrase recommender algorithm, except that the relevant list is a list of queries instead of phrases. Note that the C_M value (an internal parameter in OFSD [2]) has been set to 5 in all of the experiments.

In the first experiment, the phrase recommender algorithm based on OFSD is compared with OQD. The algorithms is applied to a part of Alta Vista 2002 query log containing 1,000,000 queries. The number of distinct queries is 496,782. The average number of items in each query is 2.06. The number of multi-item queries is 353,858. Figure 1 shows the results of applying both algorithms, where f_u=0.0008. Figure 1(a) shows the number of elements in the candidate set that is the bottleneck of the algorithm. The OFSD algorithm inspects approximately 3,000 elements for each query in the worse case, while the OQD algorithm goes over near 500,000 elements for a query. The huge difference is because of the fact that a large number of different queries are constructed by the combinations of a small number of frequent phrases. The smaller number of elements in the candidate set means the lower memory as well as the better performance. Figure 1(b) shows the percentage of queries that have been performed for an appropriate recommended phrase. Since the database is analyzed only once as a query stream, we can use the whole query log for both training and evaluation purposes. The query log is parsed sequentially. Each query Q_i is initially evaluated by the phrase recommender algorithm and then parsed as a new query by OFSD algorithm for training. In the evaluation step of Q_i, the OFSD has already been trained based on the previous queries. Q_i is considered as a satisfied query if a multi-element phrase of Q_i would be a conceptual frequent phrase in the candidate set. Therefore, the number of the satisfied queries is incremented by 1, whereas this only happens in the OQD algorithm when Q_i is a frequent query in *CS*. Number of queries that are supported partially by phrase recommender algorithm based on OFSD is 4 times the number of queries that are supported by OQD algorithm (see Figure 1(b)). The experiment proves on one hand the efficiency of the OFSD algorithm in reducing the size of the candidate set, and extracting the appropriate phrases suitable for recommendation on the other hand.

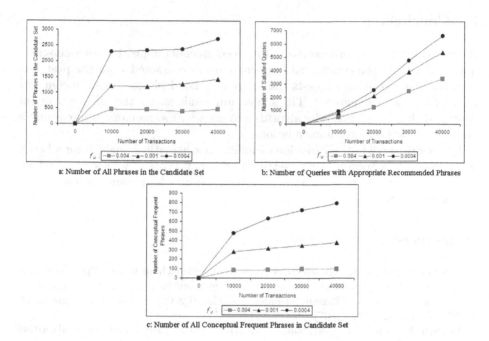

a: Number of All Phrases in the Candidate Set   b: Number of Queries with Appropriate Recommended Phrases

c: Number of All Conceptual Frequent Phrases in Candidate Set

Fig. 2. The results of applying phrase recommender algorithm to UNB query log with 45,000 queries

In the second experiment, the phrase recommender algorithm is applied to a five week query log extracted from the web site of the University of New Brunswick (UNB) web search engine (http://www.unb.ca/search.html), which consists of 45,000 queries. The special feature of the UNB query log is that its web search engine is domain specific; therefore the variety of phrases entered by users is limited. The effects of the parameter f_u to the number of phrases in the candidate set as well as the satisfied queries and CFPs are shown in Figure 2. Decreasing f_u causes more elements in the candidate set (see Figure 2(a)) while increasing the number of satisfied queries (see Figure 2(b)). Therefore, there is a trade-off between the performance of the algorithm and the query satisfaction. This shows the fact that setting f_u is an important issue to get a reasonable feedback along with a suitable performance. The percentage of satisfied queries for UNB query log is about twice of the ones in AltaVista query log while the number of input queries is less than 5%. Figure 2(c) shows that the number of CFPs does not reach a stable value for f_u=0.0004. The reason is that 45,000 queries are not sufficient for the algorithm to find a considerable number of multi-element frequent phrases. This is the reason that the number of CFPs is still increasing after 45,000 query inputs for f_u=0.0004. We intend to apply the algorithm to a larger dasaset collected from a longer period of time in the UNB web search engine.

5   Conclusions

In this paper we have proposed a phrase recommender algorithm for web search queries based on the conceptual frequent phrases extracted from the past user queries. The algorithm suggests the frequent related phrases to an incomplete query that user has entered. The simulation result shows the efficiency of the proposed phrase recommender algorithm to use very low memory and reasonable amount of successful recommendation.

Further research in this direction includes a mathematical analysis for a better understanding of appropriate f_u setting in different types of web search engines. Developing a live phrase recommender (considering the parameter LS) is also our future work.

References

1. Barouni-Ebrahimi, M., Ghorbani, A.A.: A novel approach for frequent phrase mining in web search engine query streams. In: Barouni-Ebrahimi, M. (ed.) Communication Networks and Services Research Conference (CNSR 2007), Fredericton, Canada, 14-17 May, pp. 125–132 (2007)
2. Barouni-Ebrahimi, M., Ghorbani, A.A.: An online frequency rate based algorithm for mining frequent sequences in evolving data streams. In: international conference on information technology and management (ICITM 2007), Hong Kong (2007)
3. Bast, H., Weber, I.: Type less, find more: Fast autocompletion search with a succinct index. In: Proceedings of the 29th annual international ACM SIGIR conference on Research and development in information retrieval (SIGIR 2006), New York, NY, USA (2006)
4. Charikar, M., Chen, K., Farach-Colton, M.: Finding frequent items in data streams. In: Widmayer, P., et al. (eds.) ICALP 2002. LNCS, vol. 2380, pp. 693–703. Springer, Heidelberg (2002)
5. Li, H.-F., Lee, S.-Y., Shan, M.-K.: Online mining (recently) maximal frequent itemsets over data streams. In: 15th International Workshop on Research Issues in Data Engineering: Stream Data Mining and Applications (RIDE-SDMA 2005), pp. 11–18 (2005)
6. Raghavan, V.V., Sever, H.: On the reuse of past optimal queries. In: Proceedings of the 18th annual international ACM SIGIR conference on Research and development in information retrieval (SIGIR 1995), New York, NY, USA, pp. 344–350 (1995)

Characterization of Graphs Using Degree Cores

John Healy[1], Jeannette Janssen[2], Evangelos Milios[1], and William Aiello[3]

[1] Faculty of Computer Science,
Dalhousie University, Halifax, Canada
http://users.cs.dal.ca/~{healy,eem}/
[2] Dept. of Mathematics and Statistics,
Dalhousie University, Halifax, Canada
http://www.mscs.dal.ca/~janssen/
[3] Department of Computer Science,
University of British Columbia, BC, Canada

Abstract. Generative models are often used in modeling real world graphs such as the Web graph in order to better understand the processes through which these graphs are formed. In order to determine if a graph might have been generated by a given model one must compare the features of that graph with those generated by the model. We introduce the concept of a hierarchical degree core tree as a novel way of summarizing the structure of massive graphs. The degree core of level k is the unique subgraph of minimal degree k. Hierarchical degree core trees are representations of the subgraph relationships between the components of the degree core of the graph, ranging over all possible values of k. We extract features related to the graph's local structure from these hierarchical trees. Using these features, we compare four real world graphs (a web graph, a patent citation graph, a co-authorship graph and an email graph) against a number of generative models.

1 Introduction

The primary motivation behind this paper is to compare real-life graphs such as the Web graph with various models to determine the best fit. There are many reasons for studying the structure of the Web. The most predominant of these is improving our ability to search the Web. Other research occurs in sociological analysis of communities represented by the Web graph. The ability to model the formative process underlying a real-life graph provides useful insight into its structure.

There are several methods available for comparing two graphs to determine their similarity. The most common of these approaches is a comparison of the degree distribution of two graphs. Other descriptive statistics include the distribution of clustering coefficients, the frequency of occurrence of isomorphic copies of various subgraphs [1] and the diameter. We focus on using the degree core decomposition of a graph to extract features which can be used for summarizing a graph and performing model validation.

W. Aiello et al. (Eds.): WAW 2006, LNCS 4936, pp. 137–148, 2008.
© Springer-Verlag Berlin Heidelberg 2008

A k-core of a graph is a maximal induced subgraph of minimum degree k. As we are using degree to induce these subgraphs we will often refer to them as degree cores, but may use the shorter version where the index k is relevent. It is straightforward to show [2] that the degree core is unique for a given core and a given k, and can be obtained by recursively removing all nodes with degree at most k. The degree cores of a graph can consist of multiple components. For our purposes, we generate every non-empty degree core of a graph. The components of the degree cores for all values of k form a hierarchy where two components have a parent-child relationship, when the child component is a subgraph of the parent component. We will refer to the tree thus generated as a *hierarchical degree core tree*.

To model a real world graph, we compare its hierarchical degree core tree structure to those of several generative models. More specifically, we examine the distribution of the number of components in each degree core decomposition. In our experiments we try to model four real world graphs in this manner.

2   K-cores

Cores were first introduced by Seidman [3] and popularized by Wasserman and Faust[4]. Batagelj and Zaversnik [2] generalize Seidman's work beyond simple degree to include any monotone function p. Examples of such functions range from the degree (in-degree, out-degree, directionless degree) of a vertex to the number of cycles of length k passing through a vertex. For the purposes of this paper we will use degree cores dealing with vertex degree in a similar context as Seidman used his original cores.

A k-core is the subgraph generated by recursively removing all nodes with degree smaller than k from a graph. The difference between this and simply filtering out all vertices with degree $< k$ is best illustrated by comparing their effects on a simple tree. In the case of a tree, the filtering of all degree-one nodes results in the pruning of all of a tree's leaves, whereas the degree core with $k = 2$ would prune back the leaves of a tree at each recursion, thus destroying the tree completely.

More formally, let G=(V,E) be a simple graph. A degree core is defined as:

Definition 1. *A subgraph H_k of a graph $G = (V, E)$ induced by the set $C \subseteq V$ is a k-core or a degree core of order k iff $\forall v \in C \ deg_{H_k}(v) \geq k$ and H_k is a maximal subgraph with this property.*

Batagelj and Zaversnik [2] go on to define the core number of a vertex to be the highest order of a core that contains this vertex.

The current literature indicates that degree cores have been used for variety of purposes. Most research implementing degree cores involved using them as a tool to filter the data. Our research differs from the majority of previous work in that we are interested in degree cores for their own sake, for the the insight they give into the structure of our data. Alvarez-Hamelin *et al.* [5,6] have examined the structures revealed by degree cores. Their research included an analysis of features of these degree cores across internet graphs [6] as well as the development of a visualization tool which makes use of the core numbers of a graph's vertices [5].

Some areas where degree cores have been employed in the past include:

1) Visualization: Visualization research has been done to examine the degree core structure of both real world graphs and generative models [5,6,7]. The main use of degree cores in the field of visualization is the filtering of 'unimportant' nodes, referring to nodes which have a low core number. As with most visualization techniques, the visualization research using degree cores has limited relevance to very large graphs with millions of nodes and tens of millions of edges.

2) Protein Networks: When studying protein networks, researchers are often interested in proteins that interact with other highly interactive proteins. These proteins appear in degree cores with high values of k [8,9].

3) Internet graphs: When analyzing large graphs the filtering of irrelevant nodes is often used as a pre-processing step before large graphs are analysed. Low degree nodes are often the nodes filtered when examining Autonomous System graphs of the internet. The core number of a node has been used in place of the degree for filtering these graphs [10].

4) Approximation of betweeness scores: The betweeness score reflects the number of shortest paths between all node pairs that a node lies on. This is a computationally expensive feature to calculate across a large graph. It has been shown in experiments with web graphs that the core number of a vertex is highly correlated with this score and might be used as a more efficient substitute [6].

3   Methods

In order to summarize the local behaviour of a graph, we compute every nonempty degree core of a graph and identify the connected components of these subgraphs. These components, in turn, form a hierarchy where two components have a parent-child relationship, when the latter has been immediately split from the former. We say a component, B, has split from a component A iff B is a subgraph of A and A is a component of the k-core and B a component of the $(k+1)$-core for some integer k. That is, for the hierarchical degree core tree T:

$$V(T) = \{ c_{i,k} \in CC(G_k) \mid i = (1, \ldots, \|CC(G_k)\|), i \in \mathbb{Z}, \forall k \in \mathbb{Z}^+ \}$$
$$E(T) = \{ (c_{a,k}, c_{b,k+1}) \mid V(c_{b,k+1}) \subseteq V(c_{a,k}) \}$$

Here $CC(G_k)$ is taken to be the set of connected components of graph G's degree core of order k, $c_{i,k}$ is the ith component of this degree core and $V(c_{i,k})$ is the set of vertices contained within the component $c_{i,k}$.

We refer to the tree thus generated as a hierarchical degree core tree and use this new structure for both graph summarization and feature extraction. An example of a graph and its corresponding hierarchical degree core tree can be seen in Figure 1.

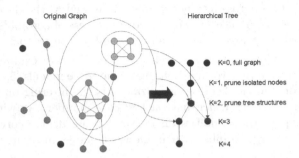

Fig. 1. A sample graph and its corresponding hierarchical degree core tree. Three components and their corresponding vertices within the hierarchical degree core tree have been circled for clarity. The larger circle is the 2-core while the union of the smaller circles makes up the 3-core.

It is clear from this definition that, given a node within the kth level of our tree, we have a representative cluster, i.e. the connected component containing the node, within the k-core of our graph. When proceeding to the $(k+1)$st layer of our tree this vertex may have:

1) no children, implying that the component it represents is no longer present within the $(k+1)$-core.

2) a single child, implying that the component remains as a single component within the $(k+1)$-core, though it may or may not have been reduced in size.

3) multiple children, implying that the removal of vertices with core number k resulted in the component splitting into multiple components.

This hierarchical tree can, at a glance, reveal a great deal of information concerning the local structure of very large graphs. It efficiently allows one to identify highly connected regions of a graph across a variety of filtering resolutions which are represented by the values of k. This tree could then be used by a domain expert to identify components of interest, similar to the way in which dendrograms are used to identify clusters of interest when performing hierarchical clustering [11].

In order to meaningfully compare massive, real-life graphs with graphs generated according to a given model, it becomes necessary to represent them as a set of features rather than as a full graph structure. The difficulty lies in determining a set of descriptive features sufficient for describing the structure of our data. Our research investigates the extraction of such features from the hierarchical degree core tree. We focus primarily of the number of components in the degree core subgraph across all k values. We also examine the size of the components in our hierarchical tree and the distribution of children between each layer of this tree. It should be noted that these are only a small subset of features that might be extracted from the hierarchical degree core tree. One might also be tempted to examine the distributional statistics such as the mean and standard deviation of component sizes at each level or the degree distribution within each of the degree core components.

4   Data

We look at a number of generative models described in the current literature [12,13] and compare these against four real world graphs in order to determine the likelihood of their being generated by one of these models. We also compare the degree cores of our various generative models to determine if we can differentiate between these families of graphs.

Our four real world datasets consist of a webcrawl of: the .gov domain [14], the NBER patent citation graph [15], an email graph [16], and a subset of the DBLP coauthorship graph [17]. Each of these graphs possesses a power law degree distribution, though they possess distinctly different structures. The web graph is a web crawl of the .gov websites from 2002 and was used as a Text REtrieval Conference (TREC) data set. It contains approximately 1.2 million unique URL's and 9.7 million links between them. The NBER patent database contains all U.S. patent citations from January 1st, 1975 to December 31 1999, consisting of approximately 3.8 million nodes and 16 million edges. The email graph is an anonymized collection representing 16 months of email sent and received by 16,000 users in the computer science department at Dalhousie University. It consists of approximately 5 million distinct email addresses and 12.7 million edges. The subset of the DBLP citation graph consists of some 15,000 non-isolated vertices and 360,000 edges. Each of the graphs consists of one major component and a number of smaller disconnected components.

The Web, along with many other real world graphs such as the four graphs above, exhibits a power law degree distribution. It has been shown that preferential attachment models mimic this and other properties of these graphs. As such, these models are ideal for the purposes of our study. Preferential attachment models are a family of models which are generally built over time, and in which the probability of a new edge connecting to a previous node is directly proportional to the degree of said node. For comparison, we also examine the degree core structure of models generated by Erdős and Rényi (ER) models [18,19]. Such models are generally referred to as $G(n,p)$ where n is the number of nodes in the graph and $p \in (0,1)$ is the probability of an edge existing between any two nodes.

The linear cord diagram (LCD) model was first proposed as a generic progressive attachment model by Barabási [20] and more rigorously defined by Bollobás [21]. In this model we start with an initial graph G_0 and add a vertex to this graph at every time step. Thus, after t time steps the graph G_t is of size $|V(G_0)| + t$. When a vertex, v_t, is added at time step t, m edges are also added connecting v_t to m vertices in G_{t-1}. These vertices are selected with probability proportional to their degree.

The main difficulty with the LCD model is that its structure is purely additive. As such, the nodes with the highest degree are simply the nodes which have been in existence for the longest time. In order to circumvent this, we examine the deletion model of Chung and Lu [22], in which there is a random probability at each step for the following changes: adding a new node connected to m previous nodes selected via preferential attachment, adding m edges whose endpoints are chosen by preferential attachment, deleting a node selected uniformly at random, or deleting m edges selected uniformly at random. Here the model is specified by the probabilities of choosing each of these options and $m \in \mathbb{Z}$.

Finally, the Chung and Lu partial duplication model [13] was examined. This model requires an initial graph for which we used K_{10} and K_{500}. Here K_i represents the complete graph of size i, or equivalently the graph of size i where an edge connects every vertex pair. At each time step, a vertex is randomly selected and copied. Each of this vertex's edges are then copied with a probability p. Chung and Lu show that the partial duplication model generates power law graphs whose power law exponents are dependent only upon the growth process, and thus independent of the initial graph.

5   Results

The hierarchical degree core trees derived from the web graph, patent citation graph, co-authorship graph and email graph can be seen in Figures 2, 3, 4 and 5 respectively. All these graphs, with the exception of the co-authorship graph, consist of a single large component and a large number of very small components which vanish completely for values of $k > 3$. In this graph, a number of components persist for $k < 22$. For clarity, we examine the degree core component distribution of the single large component of the first three graphs and the hierarchical tree for $k > 12$ in the case of the co-authorship graph.

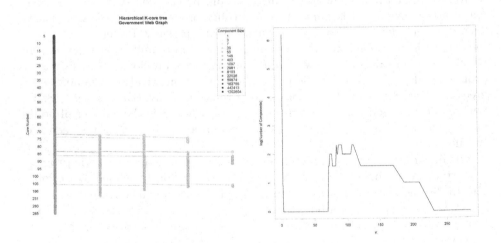

Fig. 2. The hierarchical degree core tree and corresponding component distribution for the .gov web graph. The hierarchical tree only shows the tree derived from the major component of the email graph and the component distribution is on a log_2 scale. This is due to the fact that the initial graph contains a large number of small isolated components.

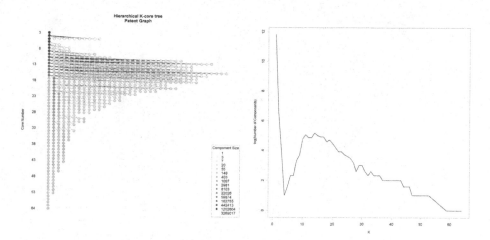

Fig. 3. The hierarchical degree core tree and corresponding component distribution for the NBER patent citation graph. The hierarchical tree only shows the tree derived from the major component of the email graph and the component distribution is on a log_2 scale. This is due to the fact that the initial graph contains a large number of small isolated components.

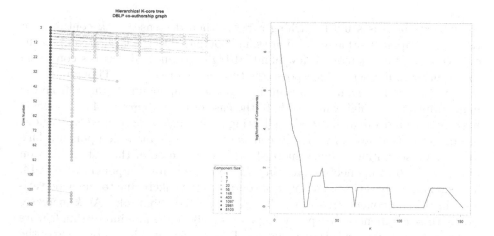

Fig. 4. The hierarchical degree core tree and corresponding component distribution for the DBLP co-authorship graph. The hierarchical tree only shows the tree derived from the major component of the email graph and the component distribution is on a log_2 scale. This is due to the fact that the initial graph contains a large number of small isolated components.

Though the hierarchical trees and the component distributions differ between these two graphs, the web graph and the patent citation graph do share one potentially interesting similarity, a unimodal component distribution. As k increases, both graphs fragment into multiple components. The number of

Fig. 5. The hierarchical degree core tree and corresponding component distribution for the Dalhousie computer science email graph. The hierarchical tree only shows the tree derived from the major component of the email graph and the component distribution is on a log_2 scale. This is due to the fact that the initial graph contains a large number of small isolated components.

components increases until reaching a peak, then decreases as k continues to rise. The components that split off are small compared to the largest component, only consisting of, at most, a few hundred highly connected vertices compared to the main component which possesses thousands of vertices. Though smaller in size, these nodes remain significant in that the components fragment from the main graph at very high values of k. In the case of the web graph, 4 components have fragmented off by $k = 83$ indicating large highly connected subregions of our graph that are only loosely connected with our main component. This unimodal distribution is much more striking in the case of the patent citation graph, with 36 components separating off from the main component at $k = 14$. The smooth decrease in the number of components is likely due to the wide variance in the component sizes of our highly connected subgraphs. As k increases slowly these components are pruned equally slowly. The more interesting feature is the smooth growth in this distribution. This is a feature shared by none of the generative models examined up to this point. This unimodal component distribution points towards interesting structures occurring near the modes of these distributions. In both the patent and citation graphs the vast majority of components fragment directly from the main component and then vanish without splitting themselves.

This strongly implies that there is a great deal of local structure within both of these real world graphs. Further, it implies that the local regions of these two graphs behave in a similar fashion, breaking apart from the main component around a single value of k. This implies that the behaviour of our vertices is far from uniform and that there is an unexplored underlying process at work.

The co-authorship graph shows striking differences to the patent and citation graphs. Firstly, its initial components are better connected and remain in the hierarchical tree to a much greater depth. Secondly, there are two components in the tree across the vast majority of the levels, occasionally vanishing to be replaced by a new child of the large component. As before, the secondary component is much smaller than the major component (consisting of less than one hundred vertices) and fragments directly from the major component.

The main component of the email graph has a different structure entirely, as seen in Figure 5. With the exception of a tiny component early on, the single large component fails to fragment at any point along the degree core hierarchy. Though this differs strongly from our previous three real world graphs, it does bear a striking resemblance to the majority of the generative models examined in this paper.

When we compare the distributions of our first three real world graphs against those of our generative models we see a clear difference. All of the models examined generate single-component graphs that vanish after reaching a given k instead of breaking apart as do most of our real world graphs. The LCD model, differs most significantly in structure, as no nodes are pruned until a given k is reached and then it vanishes entirely. This behaviour is unsurprising as it is predicted by Theorem 1.

Theorem 1. *For any graph, G, generated by the LCD model, the k-core $H_k = G$ (and is made up of a single component) for $k \leq m$ and $V(H_k) = \emptyset$ for $k > m$, where m is the number of edges added to the graph at each step.*

Proof. At each step of the iterative construction process, a vertex is added along with m edges connected to vertices already present in the graph. This process generates a single component graph with a minimum vertex degree of m. Therefore, for $k \leq m$ no vertices will be pruned, resulting in a single component.

For $k > m$ the most recently added vertex is guaranteed to have degree $= m$ and is thus removed. This pruning guarantees that the previously added vertex will now have degree $= m$. This iterative argument continues until all vertices are pruned, resulting in zero components.

In the case of the ER graph, we see a single component that persists across a very small range of k values before vanishing. Though not guaranteed, this is not surprising given the homogeneous nature of the vertices in an Erdős-Rényi graph. This difference is not entirely unexpected, as the ER graph possesses an entirely different degree distribution from our real world graphs. This analysis was included as a comparative baseline.

It appears that the CL-Deletion model is too similar to the LCD model to generate the localized behaviour which we see in our real world graphs. The addition and deletion of nodes and edges in this model occur uniformly at random across the graph. As such, they do not serve to add sufficient local structure to mimic the component distributions found in our real world examples. They do, however, serve to elongate our hierarchical tree structure by creating several distinct degree core graphs as k varies.

The duplication model was examined with the intent that it might contain more local structure. Every simulation of this model resulted in a single large component and a number of very small isolated vertices. In each case, the single main component remained in our tree across a large range of k, never generating more than a single child at each step. Though the duplication model did not match our first three real world graphs, its distribution was the most similar to that of our email graph. Both hierarchical degree core trees possessed a single large component slowly shrinking across a wide range of k values and never fragmenting.

6 Conclusion

Matching any number of features possessed by two graphs should not be sufficient to determine if both graphs were generated by the same family of models. However, failing to match features is sufficient for us to reject the hypothesis that a particular model generated a graph. Very few strong independent features have been proposed to characterize the structure of a graph. We have shown that hierarchical degree core trees possess a number of features which are useful in identifying the local structures within a graph. These features could be used along with those currently in the literature to increase the confidence in matching a graph with a generative model.

The degree core component distribution illustrates a rich local structure contained within our real world graphs. This along with the failure of our generative models to demonstrated any form of complex structure under this distribution supports the effectiveness of hierarchical degree core trees. Though in this case inspection seems sufficient to discriminate between our models and our real world graphs the next necessary step is to propose a quantitative similarity measure between our hierarchical trees. This measure might then be compared with a similar measure across the underlying graphs.

The email graph seems to match the simple structure demonstrated by the generative models, particularly that of the duplication model. One should be careful to refrain from making the assumption that the process underlying the email graph was a partial duplication process. As seen with power law models, there are many different models that could be used to generate a graph with a particular feature. It is necessary to examine a larger number of features before making any claims about the underlying process of a given graph.

The hierarchical degree core trees introduced in this paper effectively summarize the local structure contained within our real world graphs, which allows easy identification of interesting structures contained within a given graph. Even for very large graphs these trees are easily visualized. They can assist a user to determine the k value for which the degree core subgraph will contain the richest amount of information. This technique can be useful for both filtering and data exploration. In the case of at least two of our real world graphs, a good initial filter might be selecting the mode of the unimodal component distribution.

7   Future Work

Future research will involve enriching our hierarchical trees with extra information to provide a more complete summary of the graph in question. Some methods we have begun to examine include colouring vertices by the size of the component it represents or by examining the number (or proportion) of vertices of interest contained within the given component. We will continue to examine other features derived from our hierarchical trees which may prove useful in the characterization of large graphs.

Batagelj and Zaveršnik[2] extend the concept of degree cores to vertex features besides degree, such as the number of cycles passing through a given vertex. Using these other features to induce our hierarchical degree core trees may provide greater insight into the structure of our graphs.

The segmentation of our graphs into separate highly connected subgraphs suggests natural clusters within the graphs. Though these subgraphs are too small to represent major clusters in themselves, they could easily represent the backbone of larger, more loosely connected clusters. These components may or may not possess a meaningful interpretation. If they do, we will examine the use of this technique for graph clustering.

Improved graph layout techniques should improve the interpretability of our larger hierarchical trees.

Currently the code in use for computing our hierarchical degree core trees is written in C++ and makes use of the Boost Graph Library [23]. As such, it scales well to graphs containing millions of vertices and tens of millions of edges on conventional hardware. For analyzing larger graphs it will become necessary to either make use of parallel BGL or to write our own external memory algorithm.

References

1. Pržulj, N., Corneil, D.G., Jurisica, I.: Modeling interactome: scale-free or geometric? Bioinformatics 20(18), 3508–3515
2. Batagelj, V.: Zaveršnik, M.: Generalized Cores. ArXiv Computer Science e-prints (2002)
3. Seidman, S.B.: Network structure and minimum degree. Social Networks, 269–287 (1983)
4. Wasserman, S., Faust, K.: Social network analysis: Methods and applications. Cambridge University Press, Cambridge (1994)
5. Alvarez-Hamelin, I., Dall'Asta, L., Barrat, A., Vespignani, A.: DELIS-TR-0166 - k-core decomposition: A tool for the visualization of large scale networks. techreport 0166, DELIS – Dynamically Evolving Large-scale Information Systems (2004)
6. Alvarez-Hamelin, I., Barrat, A., Dall'Asta, L., Vespignani, A.: k-core decomposition: A tool for the analysis of large scale internet graphs. Computer Science, cs.NI/0511007 (2005)
7. Batagelj, V., Mrvar, A.: Pajek - Analysis and Visualization of Large Networks, vol. 2265 (January 2002)
8. Wuchty, S., Almaas, E.: Peeling the yeast protein network. Proteomics (2005)

9. Bader, G.D., Hogue, C.W.: An automated method for finding molecular complexes in large protein interaction networks. BMC Bioinformatics 4, 2 (2003)
10. Gaertler, M., Patrignani, M.: DELIS-TR-0003 - dynamic analysis of the autonomous system graph. In: Proceedings 0003, 2004, IPS 2004 (Inter-Domain Performance and Simulation) (2004)
11. Hastie, T., Tibshirani, R., Friedman, J.: The Elements of Statistical Learning: Data Mining, Inference, and Prediction. Springer, Heidelberg (2001)
12. Bonato, A.: A survey of models of the web graph. In: Proceedings of the Workshop on Combinatorial and Algorithmic Aspects of Networking (CANN 2004), Springer, Heidelberg (2004)
13. Chung, F., Lu, L., Dewey, T.G., Galas, D.J.: Duplication models for biological networks. Journal of Computational Biology 10(5), 677–687 (2003)
14. TREC: The.gov test collection (last accessed August 28, 2006)
15. Hall, B.H., Jaffe, A.B., Trajtenberg, M.: The nber patent citation data file: Lessons, insights and methodological tools. NBER Working Papers 8498, National Bureau of Economic Research, Inc (October 2001) (last accessed August, 30 2006), http://ideas.repec.org/p/nbr/nberwo/8498.html
16. Wan, X., Janssen, J., Kalyaniwalla, N., Milios, E.: Statistical analysis of dynamic graphs. In: Proceedings of AISB 2006: Adaptation in Artificial and Biological Systems, vol. 3, pp. 176–179 (2006)
17. Angelova, R., Weikum, G.: Graph-based text classification: Learn from your neighbors. In: 29th Annual International ACM SIGIR Conference on Research & Development on Information Retrieval (SIGIR 2006), Association for Computing Machinery (ACM), pp. 485–492. ACM, New York (2006)
18. Erdős, P., Rényi, A.: On random graphs. Publicationes Mathematicae (1959)
19. Erdős, P., Rényi, A.: On the evolution of random graphs. Publ. Math. Inst. Hungar. Acad. Sci (1961)
20. Barabási, A.L., Albert, R.: Emergence of scaling in random networks. Science 286, 509–512 (1999)
21. Bollobás, B., Riordan, O., Spencer, J., Tusnády, G.: The degree sequence of a scale-free random graph process. Random Structures Algorithms (2001)
22. Chung, F., Lu, L.: Coupling online and offline analyses for random power law graphs. In: Internet Mathematics (2003)
23. Siek, J.G., Lee, L.Q., Lumsdaine, A.: The boost graph library: User guide and reference manual. Addison-Wesley Longman Publishing Co., Inc., Boston (2002)

Web Structure Mining by Isolated Stars

Yushi Uno[1], Yoshinobu Ota, and Akio Uemichi

Department of Mathematics and Information Sciences, Graduate School of Science,
Osaka Prefecture University, Sakai 599-8531, Japan
uno@mi.s.osakafu-u.ac.jp

Abstract. The link structure of the Web is generally viewed as the webgraph, and web structure mining is a research area that mainly aims to find hidden communities in the Web by focusing on the webgraph. In this paper, we identify a common frequent substructure by observing the webgraph, and newly define it as an isolated star (i-star). We propose an efficient enumeration algorithm of i-stars, and try structure mining by enumerating them from the real web data. As a result, we observed that most of i-stars correspond to index structures in single domains, while some of them are verified to stand for useful communities, which implies the validity of i-stars as candidate substructure for structure mining. We also suggest that the notion of i-star can be a helpful tool for preprocessing the webgraph to have its succinct representation for further structure mining.

Keywords: isolated star, link analysis, scale-freeness, web community, webgraph, web structure mining.

1 Introduction

In the explosively evolving Web, by regarding the Web as a huge database, it is extremely important not only to obtain primary information but to find hidden information that cannot be found by naive retrievals. It is often called 'web mining', and web structure mining aims to find hidden communities that share common interests in specified topics in the Web, etc. [1,3,5,7], by focusing on the webgraph that represents the link structure among web pages.

On this model, a set of web pages of a community or its core is usually supposed to constitute a dense subgraph or a frequent inherent substructure in the webgraph, and web structure mining is actually realized by extracting them from the webgraph.

Related Works: As for significant substructures as communities, Kleinberg's hub-authority biclique model [5] is well known and attractive. Some experimental research for this direction try to enumerate (a subset of) bicliques from the webgraph and are successful for mining communities (or their cores) [6,7,8]. However, since there exist potentially enormous number of bicliques, it has become quite hard to carry out an exhaustive enumeration and to have effective outcome in the recent Web [11]. Another direction is a max-flow (or min-cut) approach that finds small cuts separating a specific set of seed pages [3]. However, this also has a drawback in the sense that we need specify seed pages in advance. On the other hand, a substructure called isolated clique [4], which has an efficient way for enumeration, can find communities or menu structures from the entire Web [11].

W. Aiello et al. (Eds.): WAW 2006, LNCS 4936, pp. 149–156, 2008.

Our Contributions: In the light of these preceding research, our contributions in this paper are summarized as follows: (i) identify a typical frequent substructure in the real (undirected) webgraph, (ii) define those structures as isolated stars so that they become easy to be enumerated, (iii) design an efficient algorithm for enumerating isolated stars that runs in (input) linear time, and (iv) exhaustively enumerated isolated stars from the real web data and give semantic analyses, which bring rich observations.

2   The Webgraph and the Web Data

The *webgraph* is a directed graph whose nodes and arcs are (web) pages and (hyper)links among pages, respectively [2]. One of the most important properties of the webgraph is its *scale-freeness*, which implies that the (in-)degree distribution of nodes shows the power-law.

For our experiments, we prepare a webgraph from the data collected by WebBase Project [10]. Here, since the graph constructed from the original web data is not necessarily simple, we apply two preprocesses: 1. remove loops, and 2. identify multiple arcs with a single arc. Hereafter, we call the webgraph constructed in this way 'the' webgraph. Table 1 shows the information of the acquired web data and the webgraph.

We use the term *domain* as a set of pages that forms a physical partition of the Web with the same string *** in the URL notation http://***/, while we consider that a *site* forms a semantic partition of the Web. We call the webgraph induced by all the nodes in a single domain the webgraph of the domain. In this paper, since isolated stars will be defined on undirected graphs, we need regard to the webgraph as an undirected graph. There may be two simple al-

Table 1. The acquired web data and the constructed webgraph

	webgraph	(undirected)
Host, Port#	WB1, 7008	
time of collection	Aug., 2003	
#domains	59,565	
#pages (nodes)	95,821,917	
#links (arcs)	1,737,732,518	345,699,858
intra-domain	1,591,587,293	345,514,732
inter-domain	146,145,225	185,126

ternatives to do it; (a) regard every arc as an undirected edge, or (b) regard bidirectional arcs as a single edge. From a viewpoint that mutual links can have significant information in the Web [11], in this paper, we introduce a webgraph that consists only of mutual links as single undirected edges by discarding all the one-way links. We refer to this webgraph as the *undirected webgraph*.

From the webgraph constructed in this way, we can observe several interesting facts, some of which are not known so far; both in- and out-degree distributions are verified to show scale-freeness with different scaling exponent; more than 1/3 of links in the original graph also have links of the reverse direction (i.e., mutual links) and that more than 99% of mutual links exist in single domains. From these, mutual links between different domains and

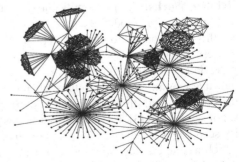

Fig. 1. A part of the "undirected" webgraph

the structures that contain those links can be expected to have significant information from the viewpoint of structure mining [1,7].

By observing a part of the 'undirected webgraph' (Fig. 1), we notice the existence of some specific substructures. One is a clique-like structure, defined as a clique or an isolated clique to facilitate their enumeration [4], and they can find useful information in the Web [11]. Another is a star-like structure, a set of nodes emanating from one central node. This structure is not necessarily dense, however, since it appears so frequently, we can expect that it has some significant information in the Web.

3   Isolated Stars

In this section, to capture star-like substructures that can frequently be observed in the real webgraph, we first give the notion of stars. There may be several possible variants that define star-like structures, of course, we especially give a definition as isolated stars in conjunction with the preceding notion of isolated cliques [4]. We investigate properties of isolated stars, and then present a naive but efficient enumeration algorithm.

3.1   Definition of Isolated Stars

A *star graph* is a bipartite graph $K_{1,|R|}$ with its partite sets $\{c\}$ and R ($\neq \emptyset$) ($\{c\} \cap R = \emptyset$). We call a node c a *center node* and a node $v \in R$ a *satellite node* of a star.

For an undirected graph $G = (V, E)$ and a subset $S \subseteq V$, if a subgraph $G[S]$ induced by S is a star graph, we call S a *star*. We sometimes denote a star $S = \{c\} \cup R$ by $S_{c,R}$, and the *size* of a

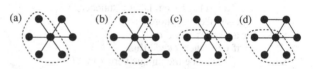

Fig. 2. (a) An isolated star, and (b), (c), (d) not isolated stars

star $S_{c,R}$ is $|R| + 1$. A *proper star* is a star $S = S_{c,R}$ that satisfies $\deg(v) = 1, \forall v \in R$ (leaf constraint). A proper star S is *maximal* if there is no proper star S' such that $S \subset S'$. For subsets X, Y of V ($X \cap Y = \emptyset$), let an edge set $E(X, Y) = \{\{x, y\} \mid \{x, y\} \in E, x \in X, y \in Y\}$. Now, if a star S of $G = (V, E)$ is (i) a maximal proper star $S_{c,R}$ and (ii) $|E(\{c\}, V-S)| < |R|$ (isolatedness), we call it an *isolated star* (*i-star*) (Fig. 2).

According to this definition of i-stars, we can easily show the following property:

Proposition 1. *For two i-stars $S_i = S_{c_i,R_i}$ and $S_j = S_{c_j,R_j}$, $S_i \cap S_j \neq \emptyset$ holds only if $R_i = \{c_j\}$ and $R_j = \{c_i\}$.*

This implies that more than one i-star can seldom share their nodes, and it makes their enumeration to be easy. This is also convenient and pertinent in the sense that i-stars can stand for disjoint communities or their cores in the Web.

3.2   Properties of Isolated Stars and Isolated Cliques

An *isolated clique* (i-clique) of an undirected graph $G = (V, E)$ is defined as a set of vertices C ($\subseteq V$) that forms a clique satisfying $|E(C, V - C)| < |C|$, and a linear time algorithm for their enumeration has been proposed [4].

As for the disjointness of i-cliques, we can show the following fact:

Proposition 2. *For two i-cliques C_1 and C_2, if they share q nodes ($q \leq \min\{|C_1|, |C_2|\}$), then $|C_1| = |C_2|$ and either $q = 1$ or $q = |C_1| - 1$ $(= |C_2| - 1)$ holds. In both cases, $E(C_1 \cup C_2, V - (C_1 \cup C_2)) = \emptyset$.*

Furthermore, we can say about the disjointness between an i-star and an i-clique:

Proposition 3. *For an i-star $S = S_{c,R}$ and an i-clique C,*
  (i) if $|C| \geq 3$, then $S \cap C = \emptyset$,
  (ii) if $|C| = 2$, then $S \cap C \neq \emptyset$ only if $|S| = 2$ or 3. In this case, $C \subseteq S$ holds.

As we saw in the series of propositions above, our definition of i-stars is given so that they are disjoint from both i-stars themselves and i-cliques except for few restricted cases. From a semantic point of view, this promises that all the i-stars and i-cliques can simultaneously stand for independent communities or their cores in the Web.

3.3 Enumeration of Isolated Stars

Once we have the definition of i-star and know that two i-stars never have an intersection except for the case that their sizes are both two, it is not so difficult to design an efficient algorithm for their enumeration. We now present such an algorithm as I_STAR below.

```
Algorithm I_STAR (Input: G = (V, E), Output: All i-stars)
S1:     for all vertex v ∈ V do
            if deg(v) = 1 then label v "unchecked";
            else label v "ignored";
S2:     for all vertex v ∈ V do
            if v is "unchecked" then
S3:             let w be the unique vertex in N(v);
                U := ∅;  i := 0;    // |U| = i
                for all vertex u ∈ N(w)
                    if deg(u) = 1 then
                        U := U ∪ {u}; i := i + 1;
                        label u "checked";
S4:             if deg(w) < 2i then     // |E({w}, V − (U ∪ {w}))| < |U|
                    output U ∪ {w} as S_{w,U} = U ∪ {w};
                    label w "checked";
```

Proposition 1 ensures the following statement.

Theorem 1. Algorithm I_STAR *enumerates all the i-stars of a given undirected graph $G = (V, E)$ in $O(|V| + |E|)$ time.*

4   Structure Mining by Enumerating Isolated Stars

In the preceding sections, we observed star-like topologies in the undirected webgraph, and regard them as a candidate substructure for web structure mining. We gave them a definition as i-stars, and designed a simple algorithm for their enumeration. A linear time algorithm for enumerating i-stars is expected to work in practice even for the entire webgraph. In fact, it can work for our webgraph of approximately 0.1 billion nodes and 17 billion links in a practical time.

4.1 Summary of the Web Data and Experiments

Since i-stars of size 2 simply imply mutual links between two pages, we neglect them as communities and focus only on i-stars of size ≥ 3 in the subsequent experiments. Fortunately, this assumption ensures that our experiments can find independent sets of nodes as candidates of communities or their cores, according to Proposition 1.

In Experiment 1, we pick 32 domains from the Web, and enumerate i-stars in the undirected webgraph of each domain. In Experiment 2, we enumerate all the i-stars from the undirected webgraph of the entire Web in Table 1, and also observe the distribution of their sizes. Here, we classify those i-stars into two categories: *intra-domain i-star* whose nodes belong to a single domain, and *inter-domain i-star* whose nodes appear in multiple domains. We can expect that most of useful communities correspond to inter-domain set of pages [1,7].

4.2 Experiment 1

Among 32 domains we examined, we focus on two domains www.gnu.org (the site of GNU Operating System) and www.keio.ac.jp (the site of Keio Univ., Japan) that remarkably showed common (topological) properties that many domains had (Table 2). Fig. 3 shows the distributions of i-clique sizes. By observing the distributions of i-star sizes, there found 88 and 8 i-stars in domains www.gnu.org and www.keio.ac.jp, respectively. Among them, we could also find some extremely large i-stars such as sizes 2,446 and 73, in domains www.gnu.org and www.keio.ac.jp, respectively.

Table 2. Data of domains www.gnu.org and www.keio.ac.jp

	webgraph		undirected
domain	#pages	#links	#edges
www.gnu.org	15,901	96,347	8,965
www.keio.ac.jp	587	5,639	1,238

Here, we examine the i-star of size 33 in domain www.keio.ac.jp, as an example. We confirmed that this i-star corresponds to a set of 'top news' pages of Keio University (Table 3). The center node of the i-star corresponds to a page that has a list of all the news contents with links to them, and we call such a page an *index page*. On the other hand, the satellite nodes correspond to pages of each news contents (*content page*), and they always had a backward link to the index page. This implies

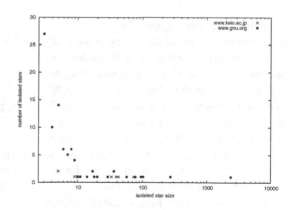

Fig. 3. Distributions of i-star sizes in domains www.gnu.org and www.keio.ac.jp

that there are bidirectional links between every content page and the index page, and they form a star, which is isolated in this case.

In addition to the above example, we verified all the i-stars of size ≥ 10 in both domains (16 and 5 i-stars in domains `www.gnu.org` and `www.keio.ac.jp`, respectively). The result showed that each of them corresponds to a set of pages consist of an index page and their content pages, and we name this kind of structure of pages an *index structure*. We could observe that all the i-stars of sizes ≥ 10 always formed index structures with no exceptions.

Table 3. URLs of the pages of an i-star of size 33 in domain `www.keio.ac.jp` that form an index structure; (a) the index page, and (b) content pages

(a)	`http://www.keio.ac.jp/news/index-en.html`
	`http://www.keio.ac.jp/news/021211e.html`
	`http://www.keio.ac.jp/news/020729e.html`
(b)	`http://www.keio.ac.jp/news/020705e.html`
	...
	`http://www.keio.ac.jp/news/010405e.html`

4.3   Experiment 2

We enumerated all the i-stars in the undirected webgraph of the entire Web shown in Table 1, and Fig. 4 shows the distributions of i-star sizes. In this figure, 'all' represents all of i-stars (including both intra- and inter-domain ones), and 'transversal' indicates inter-domain ones alone. Table 4 shows the number of enumerated i-stars classified by their sizes and by intra-/inter- categories.

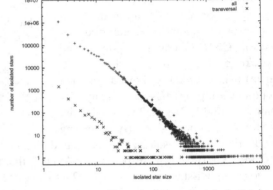

Fig. 4. Distributions of i-star and inter-domain i-star sizes

At first, we can see that the sizes of both inter-domain and all of the i-stars in the undirected webgraph roughly obey the power-law distributions. This is a quite interesting result in the sense that there is another index that shows power-law in a scale-free network. We can also find some extremely large i-stars, such as size 9,624, and this is indeed a huge index structure in a single domain (`www.shareit.com`). In contrast, the largest size of inter-domain i-star was 340.

We next verify from Table 4 that the ratio of inter-domain i-stars is approximately 0.13% of all the i-stars, which is a quite small part of all. This tells, together with the results of Experiment 1, that i-stars seem to be corresponding to index structures and are essentially inherent in single domains. However,

Table 4. Classification of isolated stars and their sizes

	size ≥ 3	any size
intra-domain	1,034,855	2,143,776
inter-domain	1,344	2,815
total	1,036,199	2,146,591

Table 5. An example of an inter-domain i-star of size 19: (a) a page corresponding to the center node, and (b) pages corresponding to the satellite nodes. They are related to libraries in Illinois.

(a)	`http://www.hccweb.com/`
	`http://www.waverly.lib.il.us/`
	`http://www.tremont.lib.il.us/`
(b)	...
	`http://www.hccweb.com/publicity/`
	`http://www.hccweb.com/faq.html`

inter-domain i-stars are expected to have different meanings from the ones that intra-domain i-stars have. Therefore, we then investigate the meanings of the pages corresponding to all of 115 inter-domain i-stars of sizes ≥ 3 lying over more than 2 domains. As a result, we observed that they are classified into the following two categories: 1. index structures, 2. sets of pages that have common interests in a specific topic, that are, candidates of communities. In Category 1, it was often the case that a single index locates each of their contents on different domains. On the other hand, in Category 2, satellite nodes of an i-star correspond to a set of top pages of some sites that share a certain topic in common, and these may be regarded as communities in the Web. Table 5 shows an example of a set of pages of an inter-domain i-star of Category 2.

4.4 Discussions

We verified through Experiments 1 and 2 that i-stars can find some candidates of communities which mainly lie over multiple domains. On the other hand, we also notice that most of the i-stars correspond to index structures in single domains, which can be viewed as site navigations or nepotistic cores [1,7]. Therefore, these index structures, especially their satellite pages, are considered no longer to be necessary for further structure mining once they are found. From this point of view, we here propose the *i-star contracted webgraph*. Remember, in preparing the i-star contracted webgraph, that Proposition 1 guarantees that i-stars of sizes greater than 2 can always be contracted in any order since they are independent.

In fact, the number of satellite nodes of i-stars of sizes ≥ 3 in the webgraph in Table 1 is 11,967,237, and therefore, if we construct the i-star contracted webgraph, the number of nodes would become 83,854,680, which is approximately 87.5% of the original one. This observation seems quite suggestive in the sense that our results not only propose a candidate structure for web mining but offer a powerful tool for preprocessing the webgraph for further utilization for web mining or give a compact representation of the webgraph which leads to a technique for compressing the webgraph [9]. We can expect to find in the i-star contracted webgraph further hidden information more efficiently that cannot be mined in the original webgraph, or at least we can say that i-star contracted webgraph can make the webgraph tractable in its size.

We already have several interesting experimental observations in these points, e.g., scale-freeness in the i-star contracted webgraph, and so on [12].

5   Conclusion

In this paper, we introduced a new graph substructure called an i-star which frequently appeared in the real webgraph. Although the definition of i-star is quite simple, it has some good properties and we can design an efficient enumeration algorithm due to its simplicity. As a result, we had a lot of useful observations by structure mining using i-stars, where it was difficult by conventional clique or biclique. We confirmed that our approach was not only successful for mining communities but can be useful for preprocessing to have succinct representation of the webgraph, and we believe that our approach can also be applied for detecting link farm spams [11,13]. It is also important to identify other characteristic substructures in the recent Web, representing such as blogs, SNS, link farm spams, and so on, for further effective structure mining. Finally, we mention that it is important to carry out these experiments on some other sets of web data.

Acknowledgments

This research was partially supported by the Scientific Grant-in-Aid from Ministry of Education, Science, Sports and Culture of Japan (Grant #15500015).

References

1. Asano, Y., Imai, H., Toyoda, M., Kitsuregawa, M.: Finding neighbor communities in the Web using inter-site graph. In: Proc. 14th Int'l Conf. on Database and Expert Syst. Appl., pp. 558–568 (2003)
2. Broder, A.Z., Kumar, S.R., Maghoul, F., Raghavan, P., Rajagopalan, S., Stata, R., Tomkins, A., Wiener, J.L.: Graph structure in the web. Computer Networks 33, 309–320 (2000)
3. Flake, G.W., Lawrence, S., Giles, C.L.: Efficient identification of web communities. In: Proc. 6th ACM Int'l Conf. on Knowl, pp. 150–160 (2000)
4. Ito, H., Iwama, K., Osumi, T.: Linear-time enumeration of isolated cliques. In: Proc. 13th Ann. European Symp. on Algorithms, pp. 119–130 (2005)
5. Kleinberg, J.: Authoritative sources in a hyperlinked environment. J. ACM 46, 604–632 (1997)
6. Kleinberg, J., Kumar, R., Raghavan, P., Rajagopalan, S., Tomkins, A.S.: The Web as a graph: measurements, models, and methods. In: Proc. 5th Int'l Comput. and Comb. Conf., pp. 1–17 (1999)
7. Kumar, R., Raghavan, P., Rajagopalan, S., Tomkins, A.S.: Trawling the Web for emerging cyber-communities. Comput. Net. 31, 1481–1493 (1999)
8. Laura, L., Leonardi, S., Millozzi, S., Meyer, U., Sibeyn, J.F.: Algorithms and experiments for the Webgraph. In: Proc. 11th Ann. Euro. Symp. on Algor., pp. 703–714 (2003)
9. Raghavan, S., Garcia-Molina, H.: Representing web graphs. In: Proc. 19th Int'l Conf. on Data Eng., pp. 405–416 (2003)
10. The Stanford WebBase Project,
    http://www-diglib.stanford.edu/testbed/doc2/WebBase/
11. Uno, Y., Ota, Y., Uemichi, A., Umano, M.: Mining communities and detecting link farms in the Web by isolated cliques. In: Proc. 2nd Int'l Conf. on Knowledge Engineering and Decision Support, pp. 179–187 (2006)
12. Uno, Y., Uemichi, A.: Structural properties of i-star contracted webgraphs. Manuscript
13. Wu, B., Davison, B.D.: Identifying link farm spam pages. In: Proc. 14th Int'l WWW Conf., pp. 820–829 (2005)

Representing and Quantifying Rank - Change for the Web Graph

Akrivi Vlachou[1,*], Michalis Vazirgiannis[1,2,**], and Klaus Berberich[3]

[1]Department of Informatics, Univ. of Economics and Business, Athens, Greece
[2]Gemo, Inria, Paris France
[3]Max-Planck-Institut für Informatik, Saarbrücken, Germany
avlachou@aueb.gr, mvazirg@aueb.gr, kberberi@mpi-inf.mpg.de

Abstract. One of the grand research and industrial challenges in recent years is efficient web search, inherently involving the issue of page
ranking. In this paper we address the issue of representing and quantifying web ranking trends as a measure of web pages. We study the
rank position of a web page among different snapshots of the web graph
and propose normalized measures of ranking trends that are comparable
among web graph snapshots of different sizes. We define the _rank change
rate (racer)_ as a measure quantifying the web graph evolution. Thereafter, we examine different ways to aggregate the rank change rates and
quantify the trends over a group of web pages. We outline the problem of
identifying highly dynamic web pages and discuss possible future work.
In our experimental evaluation we study the dynamics of web pages,
especially those highly ranked.

Keywords: PageRank, Web Graph, Web Dynamics.

1 Introduction

The Web is a highly dynamic structure that is constantly changing. The evolution of the web graph is caused by the changes in graph structure and in the web
pages' contents. One of the biggest challenges is that of efficient searching these
vast amounts of data. The research area of web search inherently involves the
issue of result ranking, since users are interested in the top results only. In this
paper we focus on the changes in the graph structure, as they predominantly
cause the changes in authority score and therefore of the web page ranking. In
particular, we aim to study the changes and the trends that appear in the rank
of a web page among different snapshots of the web graph.

We address the issue of representing and quantifying the evolution of web page
rankings –i.e. the trends that appear in their rankings– both individually and
at an aggregate level. As the different web graph snapshots may be significantly
different in terms of size we need to deal with the issue of rank normalization.

---

* Partially funded by the PENED 2003 Programme of the EU and the Greek General
  Secretariat for Research and Technology.
** Supported by the Marie Curie Intra-European Fellowship.

W. Aiello et al. (Eds.): WAW 2006, LNCS 4936, pp. 157–165, 2008.
© Springer-Verlag Berlin Heidelberg 2008

To capture the dynamism and the trends of a web page's ranking, we define the *rank change rate (racer)*. We represent the evolution of a web page through a sequence of *racer* values. Thereafter, we address the problem of finding highly dynamic pages, i.e. those with high *racer* values over a large period of time. In particular, we are interested in finding representative web pages that allow us to determine structural parts of the graph that change rapidly. Toward this goal we introduce aggregate trend measures. Finally, we outline our future work and discuss methods to identify highly dynamic pages based on *racer* values. To summarize, the key contributions of this paper are:

- We propose a rank normalization method and present *racer*, a measure that quantifies the trend of web pages through its ranking change rate.
- We discuss the problem of finding highly dynamic web pages as those that display high trends. We introduce measures to quantify the aggregate trends of a set of web pages.
- We present initial experiments of the proposed aggregate measures and estimate the expressiveness of our method to represent the evolution of web pages.

The rest of the paper is organized as follows: We give an overview of related work in Section 2, while in Section 3 we introduce the *racer* measure. In Section 4, we address the issue of estimating the dynamism of a set of web pages over time. In Section 5, we present the experimental evaluation. In Section 6 we outline our future work about methods to determine highly dynamic web pages. Finally, we conclude in Section 7.

2   Related Work

The ranking of query results in a web search-engine is an important problem and has attracted significant attention in the research community. Link-based ranking techniques like PageRank [1] or HITS [2] assess the importance of web pages based on the Web's structure. These two seminal approaches have been extended [3,4,5,6,7] and their properties have been studied intensively [8,9,10,11].

The structure of the web graph has been analyzed in different efforts showing that this graph exhibits power-law distributed degrees [12] and self-similar behavior [13]. The latter observation provides the basis for our experiments, in which we derive an understanding of the rank change rate of web pages by studying only a small subset of the web graph. The dynamics of the Web has been examined in several more recent studies. Fetterly et al. [14] put their a focus on the evolution of the Web's contents. The more recent study by by Ntoulas et al. [15] considers in addition the Web's structural evolution. Another group of related research considers the negative effects that search engines have on the Web's evolution and proposes countermeasures [16,17,18].

3   Web Page Ranking Dynamics: Rank Change Rate

In this section, we define a measure that expresses the trends of web pages' ranking among different snapshots of the web graph. Let G_{t_i} be the snapshot of the web graph created by a crawl at time t_i and let $n_{t_i} = |G_{t_i}|$ the number of web pages at time t_i. We define $rank(p, t_i)$ as a function providing the rank of a web page $p \in G_{t_i}$ according to some criterion, for example PageRank authority score values. Intuitively, an appropriate measure for web pages trends is the rank change rate between two snapshots, but as the size of the web graph may change the trend measure should be comparable across different graph sizes.

Therefore we address the need for normalization of the page ranking across graph snapshots. Assume two snapshots of the web graph G_{t_i} and G_{t_j} and times t_i, t_j respectively with $t_i < t_j$ and $n_{t_i} < n_{t_j}$. For simplicity let us assume that G_{t_i} is a subset of G_{t_j}, i.e. no nodes were removed from G_{t_i}. Let's assume that there is a page p that belongs to the web graph G_{t_i} and G_{t_j} and $rank(p, t_i) = rank(p, t_j)$, then apparently the same page is more important in the second case. For instance, assume $rank(p, t_i) = rank(p, t_j) = 5$ and $n_{t_i} = 100$, $n_{t_j} = 1000$. One would claim that the first event - page occupies the 5th out of 100 pages - is less important than the second - page occupies the 5th out of 1000 pages. Thus we motivate the definition of the normalized rank - $nrank$ - of a page in a ranked list. We impose that the $nrank$ of all pages in a ranked list sum up to 1. Thus the $nrank$ of a page p that occupies position $rank(p, t_i)$ in a list of $n_{t_i} \gg 1$ items is:

$$nrank(p, t_i) = \frac{2 * rank(p, t_i)}{n_{t_i}^2}. \tag{1}$$

We now define rank change rate ($racer$) using the normalized ranks ($nrank$) as:

$$racer(p, t_i, t_j) = \frac{nrank(p, t_i) - nrank(p, t_j)}{nrank(p, t_i)} = 1 - \frac{rank(p, t_j)}{rank(p, t_i)} * \left(\frac{n_{t_i}}{n_{t_j}}\right)^2. \tag{2}$$

Since we are interested in representing the dynamic of web pages through more than one $racer$ values, the values of different snapshots must be comparable. Thus, we define the _normalized rank change rate_ ($nracer$). In order to make the $racer$ values comparable across different graph snapshots we have to divide the $racer$ values by their value range. To determine the $racer$ value range we define the maximum value (max) as the $racer$ value when a page goes from bottom $rank(p, t_i) = n_{t_i}$ to top $rank(p, t_j) = 1$ and the minimum value (min) when a page goes from top $rank(p, t_i) = 1$ to bottom $rank(p, t_j) = n_{t_j}$. Therefore normalized rank change rate ($nracer$) for page p between graph snapshots G_{t_i} and G_{t_j} is given by:

$$nracer(p, t_i, t_j) = \frac{racer(p, t_i, t_j)}{max - min}. \tag{3}$$

Notice that we do not use footrule or Kendall's τ distance to quantify change because these measures are not sensitive to (i) the rate of change and (ii) to the relative importance of change. For example the top page falling to the 10th place and the 991st page falling to the 1000th place will be considered as events of equal importance using the footrule distance. In the Kendall's τ case reverse pairs' ranks are equally important regardless to the magnitude of the rank disagreement. An important property of *racer* is that changes in high rank positions are considered more important than changes in the lower rank positions.

4   Rank Aggregation Measures

As initial experiments show, the vast majority of the pages remains stable across graph snapshots, while significant changes in ranking are observable only for a small fraction of pages. An important problem is to identify groups of web pages that exhibit a high degree of dynamism. We first define some measurements to quantify the dynamism of a group of web pages. Based on these measures we may locate highly dynamic web pages and thereafter explore their neighborhood.

By studying the values of *nracer* we observe the trend of an individual web page over time. To capture the dynamism of a set of web pages, we have to define a measure to quantify the aggregate rank change. Notice that even if *nracer* values are calculated based on the entire graph, the proposed aggregations may consider only a subset of the graph, for example the higher ranked pages or the pages corresponding to a query result set.

A straightforward measure of the aggregate rank change value between two sets of web pages is to consider the footrule distance between the *nrank* values.

$$fracer(G_{t_i}, G_{t_j}) = \sum_{p \in G_{t_i}} |nrank(p, t_i) - nrank(p, t_j)|. \tag{4}$$

To capture the dynamics of a web graph among two snapshots but also the trend of the web pages, we define the aggregation of the *nracer* values over all pages in the transition between G_{t_i} and G_{t_j}:

$$aracer(G_{t_i}, G_{t_j}) = \frac{\sum_{p \in G_{t_i}} nracer(p, t_i, t_j)}{n_{t_i}}. \tag{5}$$

Intuitively this measure (*aracer*) represents the dominant aggregate trend in the set among the two graph snapshots. For example a positive value for *aracer* indicates that the graph generally gains importance - even though there may be pages that lose in this sense. We have to stress here that depending on the individual trends we might have a very dynamic set where the positive trends are balanced by the negative ones resulting in very small values for *aracer*.

While *aracer* represent the general trend (either positive or negative) of a set of web pages, we are also interested in the absolute dynamism of the graph in total. Thus we define *sracer* that aggregates the absolute values of the rank change rate:

$$sracer(G_{t_i}, G_{t_j}) = \frac{\sum_{p \in G_{t_i}} |nracer(p, t_i, t_j)|}{n_{t_i}}. \tag{6}$$

In our experimental evaluation we study the dynamics of the Web using two of the measures mentioned above.

5   Experimental Evaluation

In order to evaluate the effectiveness of the proposed approach, we performed initial experiments on a real dataset. The dataset is a subset of the Internet Archive obtained from its European branch[1] that contains weekly crawls of eleven U.K. governmental web sites. We constructed the web graph snapshots from this dataset, yielding a total of $560,496$ distinct nodes and $4,913,060$ edges corresponding to web pages and interconnecting hyperlinks. PageRank was computed on monthly snapshots of this graph, resulting in a total of 24 pre-computed rankings.

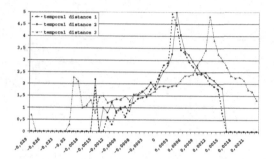

Fig. 1. The (log) frequency distribution of *nracer*

In Figure 1 we present the distribution frequency of the *nracer* values with regards to the temporal distance between graph snapshots. The plot shows the number of web pages with the particular *nracer* value in logarithmic scale. From this graph we conclude: (i) for consecutive graph snapshots, the vast majority of the pages ($80-90\%$) improve their ranking but only marginally and (ii) when the temporal distance between snapshots increases, the *nracer* values follow the same distribution but the peek is shifted to the right, thus conveying that web page rankings accumulate over time.

In the next experiment we aim to study the dynamics of web pages in the higher rank positions. We consider the subset of the top-k ranked web pages. In Figure 2 we illustrate the *sracer* values with regard to the number of pages that we consider in each subset. As expected the *sracer* values decrease as the k increases since the intersection of the two sets of pages is smaller when the set is small. This indicates that even the pages that are highly ranked change within two timestamps and verifies our assumption that the web graph is a highly dynamic structure. Thereafter, we focus on the subgraph for $k = 100$ pages.

In Figure 3, again we plot the *sracer* values but this time with regards to the position of the web page in the top-100 rank (we divide the range in bins of 10 positions each.

---

[1] An extended version of the dataset (with regard to the number of crawls) is accessible online at http://www.europarchive.org/ukgov.php

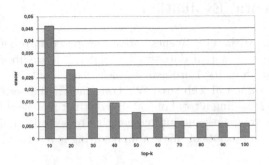

Fig. 2. *sracer* vs. size of the top-k

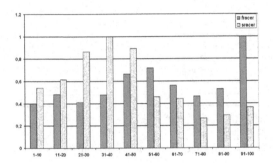

Fig. 3. *sracer* values vs. ranking position

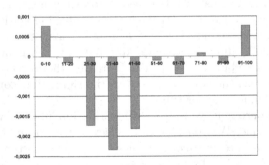

Fig. 4. *aracer* values vs. ranking position

We observe that the pages ranked between $30-40$ out of 100 have the highest *sracer* values and therefore constitute the most dynamic bin of the top-100 list.

Figure 4 presents the aggregate rank changes *aracer* vs. the ranking position. The values of *aracer* illustrate the trend of the pages. We observe that the low and high ranked pages exhibit positive *aracer* values while the mid-ranked pages exhibit negative *aracer* values.

The above experiments indicate that the trends of the web pages are not evenly distributed and that it makes sense to try to identify the most dynamic sets of pages to closely monitor them. Of course the trends and tendencies are very much dependent on the specific graph so no general rules can be extracted so far.

6 Future Work

An important problem is to identify the structural subsets of the graphs that exhibit a high degree of dynamism, i.e. high change rate values. Moreover beyond the structural parts of the Web, there is also the need for identifying topic-wise highly dynamic groups of web pages, i.e. for example based on a query term. In a first step, the issue is the identification of highly dynamic web pages which can be used as representative web pages. In a second step, these web pages may be used to determine structural parts of the web graph or to retrieve semantically related web pages and examine their dynamism. Here we deal with the first issue and discuss two methods that are appropriate to determine highly dynamic web pages.

6.1 Aggregate Ranking Across Multiple Graph Snapshots

In the previous subsection we defined measures for quantifying aggregate rank trends between only two graph snapshots. But the requirement is to be able to deal with aggregate ranking trends spanning large time periods and across many graph snapshots. Assuming a set of consecutive graph snapshots G_{t_i}, then for each pair (G_{t_i}, G_{t_i+1}) we can define a list NR_i containing the pages p in G_{t_i} and G_{t_i+1} ranked in descending $|nracer|$ score order. Apparently the top pages in NR_i represent the most dynamic ones with regards to rank change rate while the last ones in this list are the most stable ones. The objective is to aggregate the NR_i lists into a sorted list NR that best represents the pages ranked in descending order of $racer$ values over the entire time period. This problem has been extensively worked out in the past [19,20]. A straightforward solution is to consider all lists equally-weighted and handle missing values by giving them the minimum $|nracer|$ value, i.e. zero. Based on the globally sorted list NR there are two ways to choose the most dynamic web pages either (i) to choose the k web pages with the highest position in NR, where k is a fixed parameter, or (ii) to choose all web pages with aggregated score larger than a threshold value $theta$, which also is a fixed value. Afterwards, this set of pages can be used for defining highly dynamic structural parts of the Web by exploiting their neighborhood.

6.2 Pareto-Optimal Web Pages

While the aggregate of the lists NR_i returns a globally sorted list, we are interested in finding a set of highly dynamic web pages, as representative web pages. These web pages can be used to determine highly dynamic sub-graphs, based on the terms they contain or based on their locality. An appropriate set of interesting web pages is the set of the Pareto-optimal pages [21] or skyline set [22]. A web page is interesting if it is not dominated by any other, i.e. is not worse than any other in all lists. In our case, a web page over a time period is considered as highly dynamic if there does not exists any other web page that has a higher *rank change rate* in all lists NR_i. Even though Pareto-optimal web pages have not necessary the highest *rank change rate* values over a large time period, they are useful as representative web pages to determine highly dynamic set of web pages. For example consider a web page p that has a high *nracer* value in every second list $NR_{2*i}, \forall i$ and an extremely low *nracer* in $NR_{2*i+1}, \forall i$. It is obvious that web page p is highly dynamic over some time periods, but it is quite impossible for it to maintain a high rank position in the globally sorted list NR. The definition of the Pareto-optimal web pages ensure that web pages with behavior like page p are returned as highly dynamic web pages. In our future work we plan to explore their locality to identify structural parts of the web graph that are highly dynamic.

7   Conclusions

Searching in the Web inherently involves the ranking issue. Assuming PageRank as the ranking algorithm, and considering the dynamics of the Web, in this paper we address the issue of representing and quantifying the web graph evolution. Thus, we define *rank change rate (racer)*. We pose the problem of finding highly dynamic web pages and we outline our future work to enhance the applicability of the *racer* measure. We conducted initial experiments with real web data evolving over time. The results are encouraging towards a representation of the individual and aggregate ranking trends due to web graph evolution.

References

1. Page, L., Brin, S., Motwani, R., Winograd, T.: The PageRank Citation Ranking: Bringing Order to the Web. Technical report, Stanford Digital Library Technologies Project (1998)
2. Kleinberg, J.M.: Authoritative Sources in a Hyperlinked Environment. Journal of the ACM 46(5), 604–632 (1999)
3. Amitay, E., Carmel, D., Herscovici, M., Lempel, R., Soffer, A.: Trend detection through temporal link analysis. J. Am. Soc. Inf. Sci. Technol. 55(14), 1270–1281 (2004)
4. Berberich, K., Vazirgiannis, M., Weikum, G.: T-rank: Time-aware authority ranking. Internet Mathematics 2(3), 309–340 (2004)
5. Jeh, G., Widom, J.: Scaling Personalized Web Search. In: Proceedings of the twelfth international conference on World Wide Web, pp. 271–279. ACM Press, New York (2003)
6. Haveliwala, T.H.: Topic-sensitive PageRank. In: Proceedings of the eleventh international conference on World Wide Web, pp. 517–526. ACM Press, New York (2002)

7. Nie, Z., Zhang, Y., Wen, J.R., Ma, W.Y.: Object-Level Ranking: Bringing Order to Web Objects. In: WWW, pp. 567–574 (2005)
8. Bianchini, M., Gori, M., Scarselli, F.: Inside pagerank. ACM Trans. Inter. Tech. 5(1), 92–128 (2005)
9. Borodin, A., Roberts, G.O., Rosenthal, J.S., Tsaparas, P.: Link analysis ranking: algorithms, theory, and experiments. ACM Trans. Inter. Tech. 5(1), 231–297 (2005)
10. Langville, A.N., Meyer, C.: Deeper Inside PageRank. Internet Mathematics 1(3), 335–380 (2004)
11. Pandurangan, G., Raghavan, P., Upfal, E.: Using pagerank to characterize web structure. In: Proc. of Int. Conf. on Computing and Combinatorics, pp. 330–339 (2002)
12. Broder, A., Kumar, R., Maghoul, F., Raghavan, P., Rajagopalan, S., Stata, R., Tomkins, A., Wiener, J.: Graph structure in the web. Comput. Networks, 309–320 (2000)
13. Dill, S., Kumar, R., McCurley, K.S., Rajagopalan, S., Sivakumar, D., Tomkins, A.: Self-similarity in the web. In: Proc. of Int. Conf. on VLDB, pp. 69–78 (2001)
14. Fetterly, D., Manasse, M., Najork, M., Wiener, J.: A large-scale study of the evolution of web pages. In: Proc. of Int. Conf. on WWW, pp. 669–678 (2003)
15. Ntoulas, A., Cho, J., Olston, C.: What's New on the Web?: The Evolution of the Web from a Search Engine Perspective. In: Proc. of the 13th Conference on World Wide Web, pp. 1–12. ACM Press, New York (2004)
16. Cho, J., Roy, S.: Impact of Search Engines on Page Popularity. In: Proc. of the 13th Conf. on World Wide Web, pp. 20–29. ACM Press, New York (2004)
17. Cho, J., Roy, S., Adams, R.E.: Page Quality: In Search of an Unbiased Web Ranking. In: Proc. of Int. Conf. on Management of Data (SIGMOD), pp. 551–562. ACM Press, New York (2005)
18. Pandey, S., Roy, S., Olston, C., Cho, J., Chakrabarti, S.: Shuffling a Stacked Deck: The Case for Partially Randomized Ranking of Search Engine Results (2005)
19. Fagin, R., Lotem, A., Naor, M.: Optimal aggregation algorithms for middleware. In: Proc. of PODS, pp. 102–113 (2001)
20. Dwork, C., Kumar, R., Naor, M., Sivakumar, D.: Rank aggregation methods for the web. In: Proc. of Int. Conf. on WWW, pp. 613–622 (2001)
21. Kung, H.T., Luccio, F., Preparata, F.P.: On finding the maxima of a set of vectors. J. ACM 22(4), 469–476 (1975)
22. Borzsonyi, S., Kossmann, D., Stocker, K.: The skyline operator. In: Proc. of Int. Conf. ICDE, pp. 421–430 (2001)

Author Index

Lecture Notes in Computer Science

Sublibrary 3: Information Systems and Application, incl. Internet/Web and HCI

For information about Vols. 1– 4557
please contact your bookseller or Springer

Vol. 4796: M. Lew, N. Sebe, T.S. Huang, E.M. Bakker (Eds.), Human–Computer Interaction. X, 157 pages. 2007.

Vol. 4794: B. Schiele, A.K. Dey, H. Gellersen, B. de Ruyter, M. Tscheligi, R. Wichert, E. Aarts, A. Buchmann (Eds.), Ambient Intelligence. XV, 375 pages. 2007.

Vol. 4777: S. Bhalla (Ed.), Databases in Networked Information Systems. X, 329 pages. 2007.

Vol. 4761: R. Obermaisser, Y. Nah, P. Puschner, F.J. Rammig (Eds.), Software Technologies for Embedded and Ubiquitous Systems. XIV, 563 pages. 2007.

Vol. 4747: S. Džeroski, J. Struyf (Eds.), Knowledge Discovery in Inductive Databases. X, 301 pages. 2007.

Vol. 4744: Y. de Kort, W. IJsselsteijn, C. Midden, B. Eggen, B.J. Fogg (Eds.), Persuasive Technology. XIV, 316 pages. 2007.

Vol. 4740: L. Ma, M. Rauterberg, R. Nakatsu (Eds.), Entertainment Computing – ICEC 2007. XXX, 480 pages. 2007.

Vol. 4730: C. Peters, P. Clough, F.C. Gey, J. Karlgren, B. Magnini, D.W. Oard, M. de Rijke, M. Stempfhuber (Eds.), Evaluation of Multilingual and Multi-modal Information Retrieval. XXIV, 998 pages. 2007.

Vol. 4723: M. R. Berthold, J. Shawe-Taylor, N. Lavrač (Eds.), Advances in Intelligent Data Analysis VII. XIV, 380 pages. 2007.

Vol. 4721: W. Jonker, M. Petković (Eds.), Secure Data Management. X, 213 pages. 2007.

Vol. 4718: J. Hightower, B. Schiele, T. Strang (Eds.), Location- and Context-Awareness. X, 297 pages. 2007.

Vol. 4717: J. Krumm, G.D. Abowd, A. Seneviratne, T. Strang (Eds.), UbiComp 2007: Ubiquitous Computing. XIX, 520 pages. 2007.

Vol. 4715: J.M. Haake, S.F. Ochoa, A. Cechich (Eds.), Groupware: Design, Implementation, and Use. XIII, 355 pages. 2007.

Vol. 4714: G. Alonso, P. Dadam, M. Rosemann (Eds.), Business Process Management. XIII, 418 pages. 2007.

Vol. 4704: D. Barbosa, A. Bonifati, Z. Bellahsène, E. Hunt, R. Unland (Eds.), Database and XML Technologies. X, 141 pages. 2007.

Vol. 4690: Y. Ioannidis, B. Novikov, B. Rachev (Eds.), Advances in Databases and Information Systems. XIII, 377 pages. 2007.

Vol. 4675: L. Kovács, N. Fuhr, C. Meghini (Eds.), Research and Advanced Technology for Digital Libraries. XVII, 585 pages. 2007.

Vol. 4674: Y. Luo (Ed.), Cooperative Design, Visualization, and Engineering. XIII, 431 pages. 2007.

Vol. 4663: C. Baranauskas, P. Palanque, J. Abascal, S.D.J. Barbosa (Eds.), Human-Computer Interaction – INTERACT 2007, Part II. XXXIII, 735 pages. 2007.

Vol. 4662: C. Baranauskas, P. Palanque, J. Abascal, S.D.J. Barbosa (Eds.), Human-Computer Interaction – INTERACT 2007, Part I. XXXIII, 637 pages. 2007.

Vol. 4658: T. Enokido, L. Barolli, M. Takizawa (Eds.), Network-Based Information Systems. XIII, 544 pages. 2007.

Vol. 4656: M.A. Wimmer, J. Scholl, Å. Grönlund (Eds.), Electronic Government. XIV, 450 pages. 2007.

Vol. 4655: G. Psaila, R. Wagner (Eds.), E-Commerce and Web Technologies. VII, 229 pages. 2007.

Vol. 4654: I.-Y. Song, J. Eder, T.M. Nguyen (Eds.), Data Warehousing and Knowledge Discovery. XVI, 482 pages. 2007.

Vol. 4653: R. Wagner, N. Revell, G. Pernul (Eds.), Database and Expert Systems Applications. XXII, 907 pages. 2007.

Vol. 4636: G. Antoniou, U. Aßmann, C. Baroglio, S. Decker, N. Henze, P.-L. Patranjan, R. Tolksdorf (Eds.), Reasoning Web. IX, 345 pages. 2007.

Vol. 4611: J. Indulska, J. Ma, L.T. Yang, T. Ungerer, J. Cao (Eds.), Ubiquitous Intelligence and Computing. XXIII, 1257 pages. 2007.

Vol. 4607: L. Baresi, P. Fraternali, G.-J. Houben (Eds.), Web Engineering. XVI, 576 pages. 2007.

Vol. 4606: A. Pras, M. van Sinderen (Eds.), Dependable and Adaptable Networks and Services. XIV, 149 pages. 2007.

Vol. 4605: D. Papadias, D. Zhang, G. Kollios (Eds.), Advances in Spatial and Temporal Databases. X, 479 pages. 2007.

Vol. 4602: S. Barker, G.-J. Ahn (Eds.), Data and Applications Security XXI. X, 291 pages. 2007.

Vol. 4601: S. Spaccapietra, P. Atzeni, F. Fages, M.-S. Hacid, J. Kifer, J. Mylopoulos, B. Pernici, P. Shvaiko, J. Trujillo, I. Zaihrayeu (Eds.), Journal on Data Semantics IX. XV, 197 pages. 2007.

Vol. 4592: Z. Kedad, N. Lammari, E. Métais, F. Meziane, Y. Rezgui (Eds.), Natural Language Processing and Information Systems. XIV, 442 pages. 2007.

Vol. 4587: R. Cooper, J. Kennedy (Eds.), Data Management. XIII, 259 pages. 2007.

Vol. 4577: N. Sebe, Y. Liu, Y.-t. Zhuang, T.S. Huang (Eds.), Multimedia Content Analysis and Mining. XIII, 513 pages. 2007.

Vol. 4568: T. Ishida, S. R. Fussell, P. T. J. M. Vossen (Eds.), Intercultural Collaboration. XIII, 395 pages. 2007.

Vol. 4566: M.J. Dainoff (Ed.), Ergonomics and Health Aspects of Work with Computers. XVIII, 390 pages. 2007.

Vol. 4564: D. Schuler (Ed.), Online Communities and Social Computing. XVII, 520 pages. 2007.

Vol. 4563: R. Shumaker (Ed.), Virtual Reality. XXII, 762 pages. 2007.

Vol. 4561: V.G. Duffy (Ed.), Digital Human Modeling. XXIII, 1068 pages. 2007.

Vol. 4560: N. Aykin (Ed.), Usability and Internationalization, Part II. XVIII, 576 pages. 2007.

Vol. 4559: N. Aykin (Ed.), Usability and Internationalization, Part I. XVIII, 661 pages. 2007.

Vol. 4558: M.J. Smith, G. Salvendy (Eds.), Human Interface and the Management of Information, Part II. XXIII, 1162 pages. 2007.